Land Use

JOIN US ON THE INTERNET VIA WWW, GOPHER, FTP OR EMAIL:

WWW: http://www.thomson.com
GOPHER: gopher.thomson.com
FTP: ftp.thomson.com
EMAIL: findit@kiosk.thomson.com

A service of I(T)P

Land Use

The interaction of economics, ecology and hydrology

J.R. O'Callaghan
*Centre for Land Use and Water Resources Research,
University of Newcastle, UK*

CHAPMAN & HALL
London · Weinheim · New York · Tokyo · Melbourne · Madras

Published by Chapman & Hall, 2–6 Boundary Row, London SE1 8HN, UK

Chapman & Hall, 2–6 Boundary Row, London SE1 8HN, UK

Chapman & Hall GmbH, Pappelallee 3, 69469 Weinheim, Germany

Chapman & Hall USA, 115 Fifth Avenue, New York, NY 10003, USA

Chapman & Hall Japan, ITP-Japan, Kyowa Building, 3F, 2-2-1 Hirakawacho, Chiyoda-ku, Tokyo 102, Japan

Chapman & Hall Australia, 102 Dodds Street, South Melbourne, Victoria 3205, Australia

Chapman & Hall India, R. Seshadri, 32 Second Main Road, CIT East, Madras 600 035, India

First edition 1996

© 1996 J.R. O'Callaghan

Typeset in 10½/12 Sabon by J&L Composition Ltd, Filey, North Yorkshire
Printed in Great Britain by The Alden Press, Oxford

ISBN 0 412 61720 X

Apart from any fair dealing for the purposes of research or private study, or criticism or review, as permitted under the UK Copyright Designs and Patents Act, 1988, this publication may not be reproduced, stored, or transmitted, in any form or by any means, without the prior permission in writing of the publishers, or in the case of reprographic reproduction only in accordance with the terms of the licences issued by the Copyright Licensing Agency in the UK, or in accordance with the terms of licences issued by the appropriate Reproduction Rights Organization outside the UK. Enquiries concerning reproduction outside the terms stated here should be sent to the publishers at the London address printed on this page.

The publisher makes no representation, express or implied, with regard to the accuracy of the information contained in this book and cannot accept any legal responsibility or liability for any errors or omissions that may be made.

A catalogue record for this book is available from the British Library

Library of Congress Catalog Card Number: 96–84901

∞ Printed on permanent acid-free text paper, manufactured in accordance with ANSI/NISO Z39.48–1992 and ANSI/NISO Z39.48–1984 (Permanence of Paper).

Contents

List of contributors		ix
Preface		x
1	**The land use problem**	**1**
1.1	Introduction	1
1.2	Total economic value	5
1.3	The biophysical processes	7
1.4	The Natural Environment Research Council/Economic and Social Research Council land-use programme	14
	References	23
2	**The economic model**	**24**
2.1	Introduction	24
2.2	An analytical economic framework	24
2.3	Choice of empirical modelling technique	26
2.4	Data collation	30
2.5	Model construction	32
2.6	Validation	34
2.7	Model usage	36
2.8	Some alternatives to profit maximization as an objective function	37
2.9	Conclusions	39
	References	40
3	**Hydrological modelling**	**42**
3.1	Introduction	42
3.2	Hydrological processes	43
3.3	The modelling systems	45
3.4	The modelling process	57
3.5	Issues in modelling	67
3.6	Conclusions	68
	References	69
4	**Landscape ecology and land use**	**71**
4.1	Introduction	71
4.2	Processes, scale and the ecology of landscapes	72
4.3	Investigating the ecology of landscapes	75

	4.4	The development of an integrated system for evaluating the ecological consequences of land-use change at the landscape scale	86
	4.5	The application of associative modelling approaches to investigating the ecological consequences of land-use change within the context of the ecological hierarchy	90
	4.6	The application of deterministic modelling approaches to investigating the ecological consequences of land-use change within an ecological hierarchy	102
	4.7	Conclusion: whither land-use and landscape ecology?	105
		References	105
5	**Decision support systems: the NELUP example**		110
	5.1	Introduction	110
	5.2	Modelling	112
	5.3	Decision support systems	114
	5.4	Geographical information systems	115
	5.5	The NELUP spatial decision support system	118
	5.6	The database	119
	5.7	The models	121
	5.8	SDSS interface design	123
	5.9	Using the SDSS	126
	5.10	Building the DSS	127
	5.11	Assessing usability: workshops	128
	5.12	Conclusions	129
		References	130
6	**Deintensification of agriculture: an evaluation of two land-use strategies**		132
	6.1	Introduction	132
	6.2	Deintensification of agriculture in the catchment of the River Cam	134
	6.3	Deintensification of agriculture in the Pennine Dales Environmentally Sensitive Area	146
	6.4	Conclusions: land-use policy and agricultural deintensification	152
		References	154
7	**Forestry developments in the uplands and lowlands**		156
	7.1	Introduction	156
	7.2	Upland forestry: the South Tyne	158
	7.3	Lowland forestry: the Cam Basin	167
		References	177

8   The marginal value and hydrological impact of agricultural
    irrigation: a representative farm-level case study for the Cam
    river catchment   178
    8.1   Introduction   178
    8.2   Modelling approach   179
    8.3   The marginal value product of irrigated land across
          increasing yields   181
    8.4   Increasing irrigation costs to non-viable levels   183
    8.5   The hydrological impact of changes in irrigation practice   185
    8.6   Conclusions   186
    References   187

9   Discussion   188

Appendix: The catchments of the Rivers Tyne and Cam   193

Index   197

Contributors

Mr R. Adams	Research Associate – Hydrology
Mr P. Boulton	Research Associate – Ecology
Dr S. Dunn	Research Associate – Hydrology
Mrs K. Leitch	Research Group Secretary
Dr R. Mackay	Lecturer – Hydrology
Dr C. McClean	Research Associate – GIS
Mr A. Moxey	Lecturer – Economics
Mr C. Mulcahy	Research Associate – Decision Support System
Professor J.R. O'Callaghan	Director of Research Group
Mr D.R. Oglethorpe	Research Associate – Economics
Dr S.P. Rushton	Lecturer – Ecology
Dr R.A. Sanderson	Research Associate – Ecology
Dr B. White	Lecturer – Economics

Preface

The book has been produced by a research group funded by the Natural Environment Research Council (NERC) and the Economic and Social Research Council (ESRC). It is the fruits of a multidisciplinary approach to land-use studies. It brings together models in economics, ecology and hydrology and shows how they can be combined into a Decision Support System which can be used to analyse land-use policies. The research pioneers a spatial approach to problems of land use. Visit us on the Internet at http://www.cluwrr.ncl.ac.uk/nelup/nelup.html.

We are indebted to our colleagues both in the University of Newcastle upon Tyne and elsewhere who have made available to us data and progress reports on their own research. We are also grateful to them for their helpful criticism. Any errors in interpretation are entirely our responsibility.

We acknowledge the substantial contributions of the Advisory Board which has provided a strategic input to the research.

The land use problem

1.1 Introduction

Farming and forestry are practised on about 80% of the land in Britain and much of the landscape has been shaped by farmers. However, land use is no longer a simple matter of matching supply and demand for the delivery of food and fibre. There is a host of other uses, such as water gathering, recreation, wildlife conservation, which may be based on the same land. Rights of private ownership and private production coexist with less well-defined rights for public access, environmental quality and water supplies. As stated in the White Paper *This Common Inheritance* (Department of Environment, 1990), which sets out the Government's environmental policies, '... *land is the common thread. It is a finite resource and we have to find enough for all our needs – homes, jobs, shops, food, transport, fuel, building materials and recreation – while protecting what we value most in our surroundings*'.

Until the middle of the twentieth century, agriculture was practised using low input/low output production systems. Such systems had reached a steady state in recycling nutrients and were in equilibrium with the environment. The sources of pollution of air and water were almost exclusively urban and industrial. In the quarter century between 1950 and 1975 there was a technical revolution, based on petrochemicals, in the methods of agricultural production. Nitrogenous fertilizers, together with phosphate and potash, were used generously in order to relax the nutrient constraints on crop growth. A succession of new crop protection chemicals and herbicides controlled weeds and pests which improved the reliability of farming, whereas mechanization increased the productivity of labour. Plant breeders selected strains of crops with the genetic potential to convert the extra supplies of nutrients in the soil into greater quantities of biomass. Parallel improvements took place in raising the productivity of livestock for converting cereals and grass into meat and milk. The new inputs led to spectacular increases in yields per hectare. Not surprisingly, a relatively static population of consumers could not provide a market to match the gains in agricultural production and most developed countries are now faced with a surplus in the foods which can be produced from their own resources.

There are no signs that the pace of increase in yields per hectare is slackening. On the contrary, the expectations are that with the aid of molecular biology, agricultural production will become even more prolific. An unfortunate side-effect of the technical revolution was that it disturbed the stability of soil–water systems. Traditional low input/low output organic farming rotations could recycle a low level of nutrients with the minimum of leakage. Sudden applications of higher levels of fertilizers and the use of new and persistent chemicals, the basis of a high input/high output farming system, led to pollution of surface and ground waters and posed a threat to wildlife. Hedges were removed and water courses were replaced by subsurface drains in order to make it easier to operate field machinery. The public did not approve of the destruction of habitats nor of the changes in the landscape which occurred.

It seems that the requirements for land to grow food are likely to continue to decline in parallel with increasing yields of crops and livestock. The old certainty that land should be reserved for the production of food and fibre is being replaced by a diversity of interests who need land for housing, industry, amenity, recreation, water gathering, conservation and living space. Both policy makers and landowners are being encouraged to seek a range of new uses for land. EU and national policies are concerned about containing food surpluses and about protecting the environment. There is a growing demand from the public for better conservation of landscape and natural habitats. Change is also being driven by the redistribution of population, from the major urban centres to the more attractive rural regions.

Land use is also embedded in a wider debate about the future of the rural economy. The technical revolution in agriculture, together with those in transportation and communications, has brought about economic and social changes which have transformed the character of the rural economy and society. Agriculture and its related industries now rarely account for more than 15% of the employed population even in the most rural areas. Only in terms of land use is rural Britain still agricultural Britain. In all other senses – economically, occupationally, socially, culturally – it is probably misleading now to attempt to draw a sharp distinction between urban and rural economies, rather their interdependencies in the context of the national economy needs to be emphasized.

1.1.1 TRENDS IN LAND USE POLICIES

There are claims on the countryside from many sources, which are slowly reducing the importance of the production of food, while emphasizing the need for a better quality of environment and for sustainable production systems. Technological pressures to be more efficient in the production of the marketed output from agriculture point towards uniformity, whereas the revised agenda for landscape and the environment favours diversity. No generally applicable solution is on offer. With hindsight, it is clear that the policies which shaped land use during the past half century placed too much emphasis on the single objective of food production. The trend of present policies is towards encouraging a more sustainable blend of food production and countryside stewardship. The ambivalence inherent in such policies is that they do not spell out a fully integrated approach towards the development of the countryside, which could combine economic growth, employment, diversity of local potential and protection of the landscape. Decision making by landowners is increasingly circumscribed by constraints on production, whereas many of the new policy initiatives imply interdependencies between landowners and other bodies. There is a need for a medium- to long-term framework, within which policies could be analysed, and agreed. Such a framework should provide guidance for local judgements on specific developments and a mechanism for mediating conflicts in land use. Guidance plans drawn-up and agreed locally could designate what areas of landscape should be protected, the market opportunities for local agricultural production, what types of industry should be developed and the position of the vulnerable aquifers. Successful economic development needs synergy and should emphasize the interdependencies between projects.

In Britain, the Planning and Compensation Act 1991 enshrines a new plan-led system, in which the goal of sustainable development is explicit. The Development Plan System requires a District-Wide Local Plan within a strategic framework set by Structure Plans and the emerging Regional Planning Guidance. The Act specifically requires each of the planning stages to include policies on:

- conservation of the natural beauty and amenity of land;
- the improvement of the physical environment;
- the management of traffic.

English Nature, the Statutory Board which advises the Department of the Environment on issues of nature conservation lays stress on three important components of a strategy for the countryside:

- an integrated approach to rural development;
- an environmentally led approach to planning, where the environmental consequences of land use proposals can be assessed, adverse impacts minimized and opportunities taken;
- the support and enhancement of wildlife activities should be an integral part of land management.

The Countryside Commission, which advises the Government on conservation of the whole countryside and on the designation of National Parks and Areas of Outstanding Natural Beauty, expresses similar sentiments in working for 'a sustainable, multi-purpose countryside that is attractive, environmentally healthy, diverse, accessible and thriving'.

The Ministry of Agriculture, Fisheries and Food (MAFF), in its policies for agriculture, reflects the need to combine the twin objectives of producing food which meets the changing demands of consumers, and of countryside stewardship. They are encouraging the reconciliation of agricultural and environmental objectives through advice based on a better understanding of complex environmental issues, regulation to enforce minimum standards of management, and financial incentives designed to enable farmers to meet the public demand for 'unpriced' environmental goods.

The promotion of sustainable development is a cornerstone for the policies of the European Union (EU). In its Fifth Action Programme EU proposes a fundamental shift in the approach of policy making towards the environment. It wishes to achieve a balance between human activity, economic development and environmental protection by an impartial but clearly defined sharing of responsibility. The Fifth Programme focuses on the following instruments:

- Promotion of sectoral and land-use planning approaches, which take account of the environment;
- Strengthening of monitoring and knowledge on the state of the environment and the collection of related data;
- Greater use of economic and fiscal instruments;
- Strengthening technological research and development that uses resources more economically with minimal environmental impact;
- Increased transparency of decision-making processes and better understanding of problems by information, training and awareness-raising.

At the global level an integrated approach to the planning and management of land and water resources permeates AGENDA 21 (United Nations, 1993). In Chapter 10 it is proposed that:

> Integration should take place at two levels, considering, on the one hand, all environmental, social and economic factors (including, for example, impacts of the various economic and social factors on the environment and natural resources) and, on the other, all environmental and resource components together (i.e. air, water, biota, land, geological and natural resources). Integrated consideration facilitates appropriate choices and trade-offs, thus maximising sustainable productivity and use.

A new, broadly based, long-term concern for water systems is well expressed in Chapter 18 which stresses that:

> Water is needed for all aspects of life. The general objective is to make certain that adequate supplies of water of good quality are maintained for the entire population of this planet while pre-

serving the hydrological, biological and chemical functions of ecosystems.

The widespread scarcity, gradual destruction and aggravated pollution of fresh water resources in many world regions, along with the progressive encroachment of incompatible activities, demand integrated water resources planning and management. Such integration must cover all types of interrelated freshwater bodies including both surface and groundwater and duly consider water quantity and quality aspects. The multisectoral nature of water resources development in the context of socio-economic development must be recognised, as well as the multi-interest utilisation of water resources for water supply and sanitation, agriculture, industry, urban development, hydro-power generation, inland fisheries, transportation, recreation, low and flat lands management and other activities.

Integrated water resources management, including the integration of land- and water-related aspects should be carried out at the level of catchment basin or sub-basin.

1.2 Total economic value

Land, air and water may be regarded as a stock of natural resource assets which provide a range of functions, particularly a supply of food, energy and raw materials, as well as the capacity for recycling organic waste materials. A hierarchy of values, which begins by distinguishing between user and intrinsic values is shown in Figure 1.1.

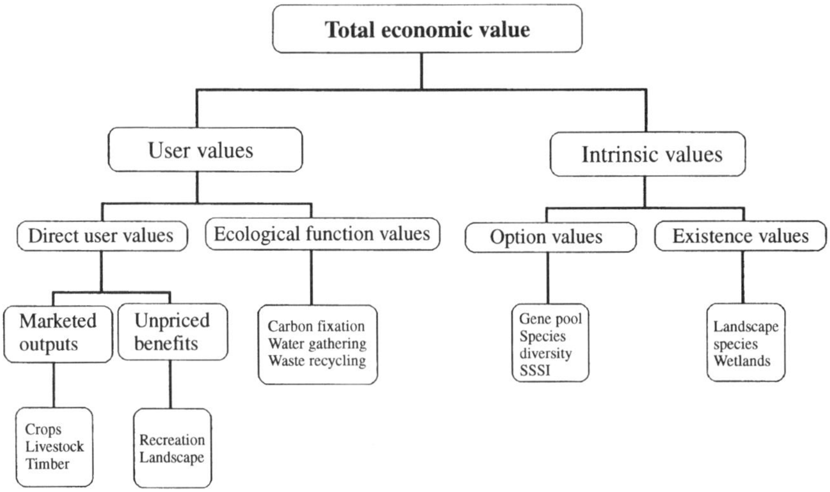

Figure 1.1 A hierarchy of values.

User values derive from the actual exploitation of natural resources, which may be classified into either direct use values or ecological function values. The most tangible of the direct use values are those for which there are markets such as crops, livestock and renewable energy. The production of commodities which are traded in well-developed markets has been comprehensively analysed using the methods of neoclassical economics. However, almost all production processes, both natural and industrial, which involve energy transformations face thermodynamic irreversibilities, which decree that not all of the energy supplied to the process can be converted into useful and final product. Consequently, some of the input becomes a waste by-product, which ultimately is rejected into the environment. What thermodynamicists see as 'rejection of energy into the environment', economists call 'externalities', and the general public know as 'pollution'. As production becomes more intensified or concentrated, the quantities of waste products become greater and tend to overwhelm the recycling capacity of the environment. The problem becomes particularly acute with discharges of chemicals, some of which are toxic. The intensification of agricultural production depends on imports of fertilizers, pesticides and animal feed onto farms, which has increased the probability of diffuse pollution from them, especially to surface and ground waters. Intensification of agricultural production, together with industrial pollution, has changed the environment for flora and fauna. Inevitably some species have been unable to cope with the changed conditions and have disappeared, many have declined, and only a few have benefited.

Whereas the majority of those who hold property rights in the countryside derive a return on their investment through the net value of the marketed outputs from their land, unpriced benefits are enjoyed collectively by the public. There are 'goods' such as landscape and wildlife, which arise from the processes of land formation and land use.

Ecological function values are the physical and chemical processes, associated with photosynthesis and decay, which support life itself. The life support system is maintained by solar energy, which drives the two major natural cycles, carbon–oxygen and hydrological. The sophisticated criterion of pollution is: does it interfere with any parts of the biological cycles on land or in water? A role of science is to give early warning of any pollutant or process which threatens these life-support systems, and to alert the public, industry and governments to take action in order to safeguard not just human health, but biota and the amenity of landscape.

Intrinsic values are less easy to define, and suggest values that reside in natural resources, and which are not associated with actual use. Option values represent a willingness to pay for the preservation of a natural resource against the probability of its use by future generations;

for example, maintaining as large a gene pool as possible, or conserving plant species, in the hope that they may yield medicines, which have yet to be identified. Existence value is a value placed on an environmental good, which is unrelated to any known actual or potential use of the resource. It refers to the satisfaction people derive from the knowledge of the existence of a species, ecosystem, lake or scenic view.

The scale of values placed by society on the different components of total economic value is not constant. In times of food scarcity, the main emphasis is on the marketed outputs. On other occasions, such as now, concern for the unpriced benefits and intrinsic values is expressed. At all times, the maintenance of life-support systems depends on the proper functioning of the biological cycles.

1.3 The biophysical processes

The biophysical processes, underpinning the 'user values' of the economists, are those which maintain life in all its forms. Solar energy is the driving force for both the hydrological cycle and the carbon–oxygen cycle. Most of the radiant energy reaching the surface of the earth is used to evaporate water. Solar heat evaporates water directly into the atmosphere from the free surfaces of the rivers, lakes and oceans. Over land the pathways are more complicated: water may be evaporated from bare soil, or, where there is a crop canopy, water may be evaporated directly into the atmosphere from the surface of a wet canopy. Usually water is withdrawn from the soil by the roots and transpired into the atmosphere through the stomata of the plants forming the canopy. A small fraction of the water, together with the nutrients dissolved in it, is retained by the plants, where it is transformed into biomass, which then becomes the source of energy for living organisms. Closely linked with the carbon–oxygen cycle are the cycles of the mineral nutrients essential for plant growth. In relation to both agricultural and environmental issues the uses of both nitrogen and phosphate fertilizers are important.

1.3.1 THE HYDROLOGICAL CYCLE

Hydrology is concerned with all aspects of the water cycle: its sources, distribution and circulation, its effects on the environment and on life in all forms. A major part of the subject is concentrated on studies of precipitation, run-off, transpiration, exchanges in the unsaturated root zone and deep percolation into the ground water.

A simplified representation of the hydrological pathways is shown in Figure 1.2. Water enters the system as precipitation in the form of rain or snow. If rain falls on a crop canopy, some of the water intercepted by the canopy is evaporated from the leaves and returned directly to the

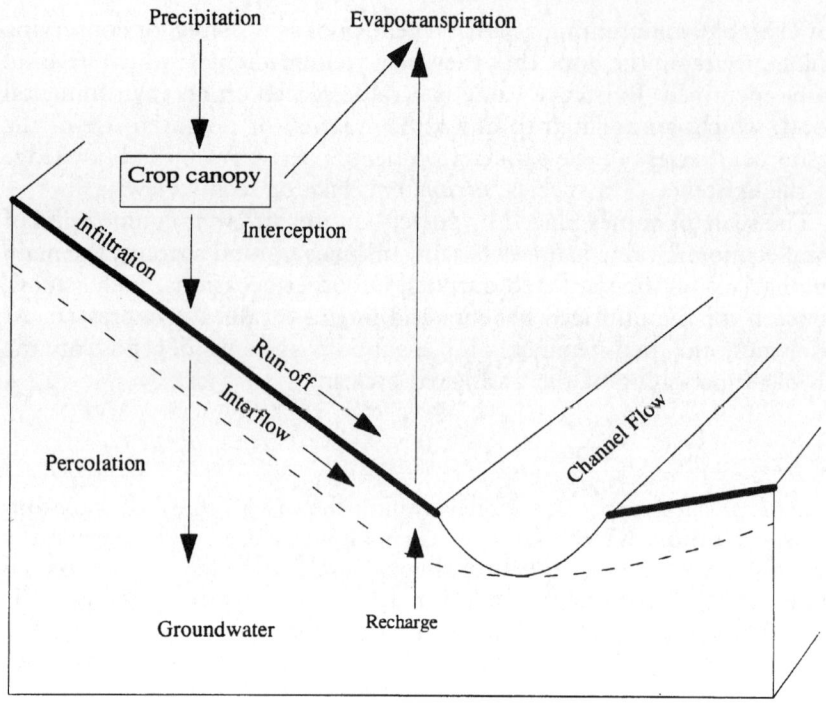

Figure 1.2 Principal hydrological pathways.

atmosphere. The remainder of the rain falls through the canopy reaching the ground. Some of it may flow along the ground, part to be evaporated and part as surface run-off to a stream or river. Topography defines the spatial domain, the catchment, within which the water must flow. The remainder infiltrates into the ground to be stored temporarily in the unsaturated zone just below the surface, from which it may be transpired by a crop or may percolate into the saturated zone below the water-table. Water in the saturated zone may flow under the soil surface and reappear as a spring to discharge into a stream or river. It may also percolate into a deep aquifer from which it may be abstracted by pumping to provide a water supply.

The network of pathways form an input–output system which describes a water-balance for a catchment or subcatchment over a chosen period of time, usually a year.

$$\text{Precipitation} = \text{Run-off} + \text{Evapotranspiration} \pm \text{Storage}$$

Water budgets are notoriously difficult to calculate. Even rainfall is difficult to estimate if its spatial variability is to be taken into account. Evapotranspiration is a very complex process involving heat and mass transfer. There are no reliable ways of measuring changes in storage which can take place at the surface, in the unsaturated and the saturated zones.

The quality of water may be affected through taking contaminants into solution as it flows over the surface of the ground or through the root zone.

1.3.2 THE CARBON–OXYGEN CYCLE

The carbon–oxygen cycle comprises two parts: the photosynthetic part in which carbon dioxide and water are converted into carbohydrates and the decomposition part in which the organic molecules are broken down again into carbon dioxide and water.

The complex process of converting solar energy into plant material can be represented in a simplistic way by the endothermic photosynthesis equation:

$$CO_2 + H_2O \rightarrow [CH_2O] + O_2$$

$[CH_2O]$ is a non-committal formula for the basic unit of a carbohydrate molecule and six of these would yield glucose ($C_6H_{12}O_6$). The process by which light energy is used in the leaf of a plant to activate the combination of two low-energy substances, carbon dioxide and water, into a high-energy form of sugar, such as glucose, is extremely complex. However, the process is the principal route by which we can capture solar energy, and, once glucose is synthesized in the leaf, it becomes the source of energy to support further complex biochemical reactions in the plant such as the production of proteins, enzymes, nucleic acids, starch and cellulose. The energy for these processes is released in a reverse form of the photosynthesis equation, where sugar is oxidized in exothermic reactions which release energy.

The other part of the carbon–oxygen cycle is one of aerobic decay which occurs in numerous biochemical steps, and through various ecological chains of decomposer and detritivore communities. When a plant dies, or the animals which feed on the plants die, the reducers or decomposers obtain a share of the energy originally fixed by the plants. Reducer populations consist mainly of bacteria, fungi and protozoa. The reducers serve an essential role in the ecosystem by recycling the carbon and mineral elements for reconversion into biomass. The energy stored in organic matter is released in oxidative reactions, in which oxygen serves as the ultimate electron acceptor. When the food chain is disrupted due to localized exhaustion of the oxygen supply or to toxicity towards the micro-organisms, the carbon–oxygen cycle is placed at risk. The capacity of micro-organisms for breaking down pollutants and rendering them harmless is prodigious, but microbial communities are a vulnerable part of the life cycle and any pollutant that threatens them is very dangerous for the whole cycle.

1.3.3 THE NITROGEN CYCLE

Nitrogen is an important element in the complex biochemical syntheses that are part of plant growth. Much of the impressive improvements in plant productivity, which have recently taken place in agriculture, are attributed to increased applications of nitrogenous fertilizer to the soil. Unfortunately some of that nitrogen was leached out of the soil, and found its way into the surface and groundwaters, where it has environmental implications for the quality of drinking waters and the growth of algae.

Although nitrogen gas constitutes almost four-fifths of the air in the lower atmosphere, it cannot be assimilated directly from the atmosphere in its gaseous form. Nitrogen is fixed from the atmosphere by three processes.

- A small amount is fixed atmospherically when lightning and ultraviolet radiation form oxides of nitrogen in the atmosphere. This is augmented by the NO_x constituents found in the exhaust gases from internal combustion engines, and by the ammonia gases released from livestock units and the soil. Some of these gases are dissolved in rain and washed into the soil.
- Some is fixed biologically by micro-organisms in the soil.
- A significant amount is fixed industrially by chemical conversion into nitrogenous fertilizers, using, for example, natural gas as a feedstock, to supply all the energy and some of the hydrogen, in a reaction which combines nitrogen from the air with steam to produce ammonia.

The flows of nitrogen into and out of the soil are shown in Figure 1.3. The problem is complicated by the demands for nitrogen from the soil microbes participating in the biodegradation processes in the root zone, which are an important part of the carbon cycle. The microbes need nitrogen for their own growth processes and are in competition with the plants for what is a scarce resource. In effect, there are two pools of nitrogen in the root zone. There is a large pool of nitrogen which is tied up in complexes with the organic matter, of the order of 5 tonnes N/ha in arable soils and 15 tonnes N/ha in old grassland. There is a very much smaller pool of available inorganic nitrogen for which the microbes and plants are in competition in order to meet their immediate requirements.

The nitrogen removed from the inorganic pool by the microbes is

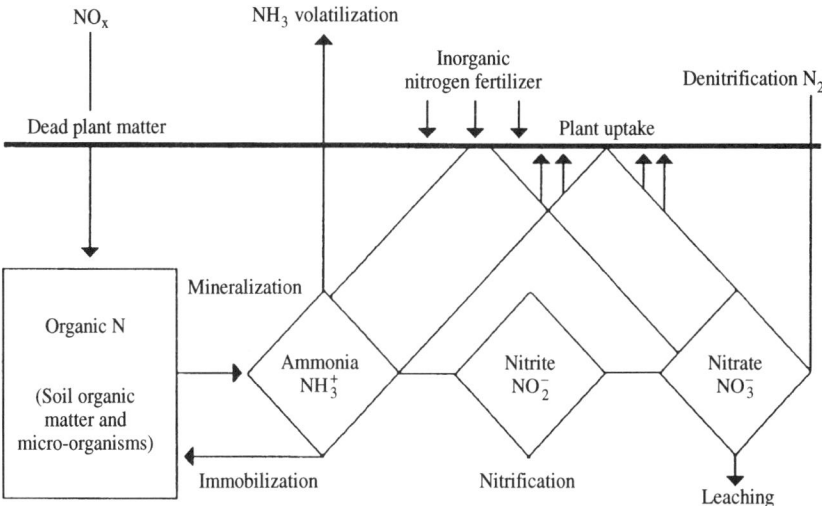

Figure 1.3 A schematic nitrogen cycle.

immobilized into the organic pool in the form of microbial cells and humus, from which it is 'mineralized' back into the nitrate pool after a number of years.

The ammonia in the inorganic pool tends to have a short life. It may either be taken up by the plant roots, or oxidized to nitrate through the microbial process of nitrification. The negatively charged nitrate ions are not absorbed by the negatively charged colloids in the soil. Unless they are taken up by the plants or microbes, they will be leached out of the root zone in the water that later reaches the streams as interflow or percolates to the aquifer.

1.3.4 NUTRIENT CYCLES

In order to grow and develop, all living organisms require a supply of raw materials which fall within three broad categories: a source of energy in the form of oxidizable organic matter, water and minerals.

Green plants differ from most other organisms in that they can fix, by photosynthesis, their own supplies of primary organic material in the form of glucose from carbon dioxide and water. The plants draw on this material for the biosynthesis of more complex organic molecules and as a source of energy which is released by respiration. Some of the glucose is converted into cellulose, a small part is combined with nitrogen to form amino acids and proteins, and much of the remainder is converted into sucrose, which is the main translocation medium of a plant.

Water fulfils several functions in plant growth. It is an active ingredient in the photosynthesis reaction, it takes into solution nutrients from the soil and conveys them to the roots, it moderates temperature fluctuations and is the major constituent by weight of the growing plant.

A total of 17 elements are needed in the processes of plant growth and development. Four (carbon, hydrogen, oxygen and nitrogen) are commonly available in the environment, and three of these (carbon, hydrogen and oxygen) account for about 90% of the dry weight of a plant and are fixed by photosynthesis. Nitrogen, although abundant in air, has to be 'fixed' for the plants either biologically or chemically. The remaining 13 elements are found in the parent rocks from which soil is derived. If they are not all present in sufficient quantity in a particular soil for optimum plant growth, deficiencies can be rectified by purchasing supplements in the form of fertilizers.

A nutrient balance of the soil in the root zone during one or several crop cycles is represented as:

$$\text{Nutrients in} = \text{Nutrients out} + \text{Nutrients stored}$$

Nutrients into the system may come from such sources as

- residues from a previous crop;
- rainfall carrying nitrogen and sulphur in solution;
- organic manures from a livestock enterprise;
- application of commerical fertilizers.

Nutrients may be removed from the system in a number of ways:

- fixed in the harvested part of the crop;
- carried away as leachate in the surface or ground water;
- removed in soil carried away by wind or water erosion;
- volatilized as a gas into the atmosphere.

The quantity of nutrients stored in the soil is not a constant quantity. It can decrease as nutrients are released from the soil in response to plant demand or to leaching and it can increase as some of the input is consumed by microbes and fixed organically in the soil or fixed chemically with clay minerals. The way in which nutrients are partitioned between plant uptake, storage and losses is difficult to estimate, since it is dependent on the climatic and husbandry conditions prevailing in a particular year. The nutrient balance is further complicated by the fact that an element may exist in several forms, and in many cases, there is not a single pathway but several, some of which are in parallel. Supply and demand are approximately balanced by supplementary additions of fertilizers, which provide nitrogen, phosphorus, potassium and occasionally sulphur. Rapid partial chemical analyses of samples of soil allow the quantities of fertilizers, which should be added to the soil in order to create the right nutritional environment for a crop, to be crudely estimated. The technique suffers from limitations of sampling in reflecting spatial variability, of estimating crop demand and of weather variations.

Excess leakage of nutrients out of the soil into the water system is a matter of great environmental concern.

1.3.5 ECOLOGY

Ecology, which is the study of biodiversity, is concerned about the interactions between organisms and their environment, and how those interactions determine the distribution of both plants and animals. An ecosystem is a network of energy and mineral flows in which populations of plants, animals and micro-organisms perform different functions, which are complementary, and integrate in a holistic way. The organisms may be assigned to three broad functions: the producers (plants), the consumers (animals) and the reducers (micro-organisms), which are essential to the operation of the biophysical processes. The plants are linked to photosynthesis and evapotranspiration, whereas the animals and micro-organisms consume, decompose and recycle both organic and inorganic materials. Ecology operates at the interface, where the carbon and hydrology cycles meet, and where water is the universal solvent that collects and transports minerals to growing plants and assists the processes of decomposition in dead organic matter. It is also where heat is absorbed in bringing about a change of phase from liquid to vapour in the hydrological cycle.

The interactions between organisms and the environment are studied at three broad levels:

- at the level of the individual organism, where research is often carried out under closely controlled laboratory conditions in order to understand how environmental factors impinge on the life cycle and welfare of a chosen organism, be it plant or animal;
- at the level of the population, being a collection of individuals of the same species at different ages and different stages of development; and
- at the level of the community of different populations, their structure and the pathways followed by energy, nutrients and other chemicals as they pass through them.

Notwithstanding the complexities of the scales, pathways and interactions in community ecology, it is the branch of the subject which is most relevant to studies in land use. The structure of the landscape needs to be perceived not only from the human perspective, as a place where agriculture and forestry are practised, water collected and scenic beauty experienced, but also with regard to the scale at which it should be organized and managed in order to maintain biodiversity. There are vast differences between the space and time scales at which humans can operate in comparison with those at which landscapes and ecosystems respond. Unfortunately, destruction is likely to be more rapid and irreversible than rehabilitation.

Broad constraints are imposed on the structure of a landscape by its geology, soils and microclimate. However, within those wide limits very different types of landscape can emerge as the result of the way in which the natural resources are managed. It is relatively easy to distinguish between man-made and natural landscapes. In the man-made landscape boundaries are sharp, agriculture is intensive, monocultures predominate and what wildlife exists is confined to discrete patches within the landscape. On the other hand natural landscapes are characterized by their blurred boundaries, where a continuous mosaic of vegetations often replaces the discrete patches.

Spatial organization is an important attribute of a landscape. Landscape ecologists emphasize three aspects of spatial organization especially in man-made landscapes:

- an understanding of the relationships between habitat patches (their composition, size and distribution in the landscape) and the kinds of viable populations that are likely to colonize them;
- the function of corridors in linking the patches into a network and the extent to which they provide connectivity and increase the permeability of the landscape;
- the notion of 'matrix' as a concept which defines the character of a landscape in a distinctive and recognizable way.

One of the challenges facing landscape ecologists is to propose ways in which the resources of the countryside should be deployed and the flows of energy, material and species be managed in order to maintain sustainable systems of land and water use.

1.4 The Natural Environment Research Council/Economic and Social Research Council Land Use Programme

The objective of the NERC/ESRC Land Use Programme (NELUP) was to provide a rationale for the way in which decisions may be reached about land use and for predicting how these decisions are likely to affect the environment, especially how they might impinge on hydrology and ecology. The research examines how ideas of total economic value may generate different spatial approaches to land use, and at the same time, check how they might affect the sustainability of the biophysical processes. There is an underlying assumption that decisions about land use are market-driven. The NELUP approach is to devise an analytical framework, which draws on the large quantities of data that are relevant to decisions that have to be taken. The data are brought together in quantitative models, which describe, in a simplified way, the principal processes involved in land use. A decision support system, based on the biophysical–economic models, is proposed and it allows decision makers to explore the likely outcomes of different policy proposals for land use. In other words, to answer at the policy-making stage, 'what-if' questions.

The research brings together a number of new technologies which make a spatial approach to land use studies feasible.

- **Remote sensing**, in combination with global positioning systems, allows large amounts of environmental data to be acquired and geo-referenced together with the opportunity of repeated runs which could build up a time-series of events. Modern methods of data acquisition, combined with telematics, provide an abundant supply of information, which can be used in the first instance to augment other sources of data on land use, particularly on the large number of factors not amenable to observation by remote sensing. The pressing problem is to analyse as much of the data as possible in order to provide a better understanding of processes and systems, from which future performance can be predicted with greater confidence, and gaps in existing knowledge identified.
- **Geographic information systems**, which provide both the hardware and software for storing, manipulating, analysing and combining spatially referenced data from several sources in forms of compatible input to simulation models. Efficient data management systems are the crucial link in the chain between acquisition and interpretation. Unfortunately much of the older data is neither geo-referenced accurately enough nor of the quality to provide information at the spatial resolution of the remote sensing systems that are operating now. Though the transformation of older data into GIS formats is time-consuming, it is a one-off entry cost worth paying in order to make the information available for improved methods of analysis.
- **Computer systems**, which have the capacity to handle the large quantities of data necessary for the description of environmental processes and to simulate their behaviour. The spatial analysis and display of information has the potential to improve severalfold our detailed perception of what is taking place in the landscape. For example, it opens up an entirely new approach to catchment planning, where traditionally measurements were concentrated

on flow in the river channel. Now the analysis can start from the distribution of the rainfall, the effects of different crops in various locations and the influence of different configurations of wetlands on the probability of downstream flooding.
- **Graphical user interfaces**, which can display complex information in interactive, user-friendly graphical ways. Spatially distributed information, which defies concise description, is conveyed clearly on a map. Computerized decision support systems (DSS) have been created to help decision makers confront poorly structured problems through direct interactions with data and analysis models. Traditionally, forecasts of policy effects have been produced by a team of specialist researchers whose findings may themselves need specialist interpretation. The underlying concept behind a DSS is that it should be used by the policy makers themselves rather than their specialist researchers.

1.4.1 THE METHODOLOGY

Realization of the general objective of the programme depends on an understanding of how the processes of land use function, both the way in which land is allocated between different activities and the likely impacts of those activities on the environment. A distinction may be drawn between highly concentrated activities, of an industrial kind, which can be treated as point sources, and diffuse activities, such as agriculture, which are spread over larger areas. Both kinds of activities may impose external costs on society through degraded water supplies and altered ecological habitats. The connections between land use, water and ecology were chosen as the important linkages in a biophysical–economic model of land use. If such a model were to be useful for decision makers, it should allow the processes and linkages to be explored spatially as well as quantitatively. As rural land use is dominated by agriculture at the river basin scale, it was essential that the model should deal with diffuse activities, while providing a facility for incorporating point sources into the analyses.

1.4.2 THE LAND COVER TRIANGLE

Decisions about the allocation of land between competing uses are influenced to a large extent by market forces, which emphasize rates of financial return and the short term. However, these decisions may have long-term implications for the landscape, for the quantity and quality of the water reaching the rivers and reservoirs, and for ecology. A common factor which may be used to link market forces, hydrology

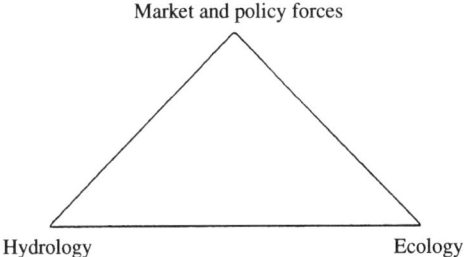

Figure 1.4 The land cover triangle.

and ecology, as shown in Figure 1.4, is land cover. When a landowner decides to use an area of land, the decision implies choosing a cover for the soil in the form of a crop, cereal, grass, forestry or even an impervious concrete surface.

The cover intercepts the rainfall, influences the amount of run-off, depth of root zone and evapotranspiration. The assemblages of plants and animals likely to be found at a particular site are strongly related to the cover at the site and in many cases define our view of it.

Three sources of land cover data, which differ in the way they are spatially referenced, have been used in the development of NELUP.

(a) Remote sensed land cover map

A remote sensed classification of land cover is produced by the Institute of Terrestrial Ecology (ITE). It is based on the classification of images captured by the Thematic Mapper sensor on the Landsat satellite. The sensor records the landscape on a 25 m pixel grid, representing different cover types together with structural patterns within the landscape. The definition of cover types and the choice of land classes is based on the experience within the ITE of interpreting data collected during surveys on the ground, from low-flying aircraft and from satellites. A total of 25 land cover classes are used in the mapping of Great Britain at a 1 km grid scale; they range from sea/estuary (class 1) through agricultural and amenity grassland (6, 7, 9) upland bog (17) and arable land (18) to urban/industrial (21) and felled forest (23). There are 18 seminatural vegetation types subdivided into three woodland classes, four heathland communities, three wetland types, seven grassland habitats and bracken.

The Land Cover Map of Great Britain has been integrated with the data of the Countryside Survey 1990 which was based on a stratified random sample of 508 1 km squares. The main objectives of the Countryside Survey were:

- to record the stock of countryside features in 1990, including information on land cover, landscape features, habitats and species;
- to determine change by reference to earlier surveys in 1978 and 1984;
- to provide a baseline of information against which future changes could be assessed.

The Land Cover Map of Great Britain in combination with the Countryside Survey offers a very comprehensive dataset for a wide spectrum of studies in land use.

(b) Agricultural census data

An agricultural census is conducted annually at the beginning of June by the Ministry of Agriculture, Fisheries and Food and is a principal source of data on agricultural land use. It provides data on crops, types of grassland and livestock numbers. Information supplied by the occupiers of individual agricultural holdings is confidential, but the Parish Summaries provide the most geographically detailed results that are available publicly. The agricultural statistics contain elements of spatial indeterminancy, from the way in which farms are allocated to a parish and from variability in land capability.

(c) Field surveys of habitats

Land cover can be observed directly by ground survey. In view of the requirement that definitions of land cover should be carefully drawn and widely accepted, well documented, standard definitions of land cover should be used. The Nature Conservancy Council Phase 1 Vegetation Survey methodology provides such a framework.

1.4.3 THE QUANTITATIVE MODELS

The quantitative models, which have been developed to describe the interlinked processes between a land use decision and its impacts on the environment, may be grouped under the headings of economics, hydrology and ecology.

(a) Regional agricultural economic model of land use (Chapter 2)

An economic model of a catchment brings together land use and land productivity. It provides understanding of how crop and livestock production systems are likely to adapt in response to changes in policies which affect land use. The model has three major components:

1. A land classification system, reflecting the spatial distribution of potential agricultural productivity. It is based on soil physical properties, topography and climate;
2. A Bayesian method of estimating technical coefficients, which relate input usage to output levels for a range of land uses. The prime source of such data is the Farm Business Survey conducted for MAFF.
3. A linear programming (LP) model of the catchment, which estimates existing intensities of crop and livestock production on the basis of historical production data, and shows how the data may be used to extrapolate new policies into the future.

(b) The hydrological models (Chapter 3)

NUARNO, a fast, macro hydrological model provides output at sub-catchment resolution, and is used to monitor discharges, as a way of detecting changes which may result from new land uses, and are worthy of more detailed analysis.

The Systéme Hydrologique Européen (SHE) hydrological model was chosen for simulating both the flow and quality of water and for predicting the hydrological consequences of land use change at a more detailed level. The SHE is a physically based, spatially distributed modelling system, well-suited to examining in detail changes in both fluxes and storage at a particular location. It calculates canopy interception, evapotranspiration, as well as overland and subsurface flows in both the unsaturated and saturated zones.

(c) The ecological models (Chapter 4)

A matrix model is used for assessing some of the ecological consequences of land use change. The model uses the ITE Land Cover Map of GB to predict the likelihood of occurrence of individual species or species-assemblages in km squares. It assumes a three-level hierarchy: the lowest level contains information about species-assemblages; the second, data on the occurrence of species-assemblages within different land-cover types and the third, the spatially referenced remote-sensed land cover map giving information on the distribution of different cover types on the land surface. Land use change is divided into two kinds: changes which result in whole cover types being transformed from one type to another, and changes in intensity of land use. Changes in land use at their simplest, are defined in terms of changes within two of the hierarchical divisions. Changes in cover result in changes at the third level of the hierarchy and changes in intensity of land use in changes at the second level, although it is possible for land use changes to cause between-level changes in the hierarchy. The matrix model is appropriate for dealing with assemblages of plants or invertebrates, which are relatively immobile in the landscape.

Prediction of the ecological consequences of changes in land use can be made either where the changes in cover are defined by the user, or as an input from the calculations of the economic model.

A within-cover change model for plant communities, based on environmental and management conditions associated with plant communities listed in the National Vegetation Classification is available to make predictions of plant community and species composition in response to changes in the intensification of agricultural management practices.

A GIS approach has been adopted for modelling the potential habitat distribution of birds and mammals which can range over the landscape in order to satisfy different aspects of a mix of habitat requirements. It uses a screening approach to habitat characteristics in order to match the known preferences of a particular species with their physical availability in the landscape.

1.4.4 THE DATABASE (CHAPTER 5)

An important contribution of NELUP to land-use studies in general is to bring together from various sources the data that are relevant to the major processes involved in land use, hydrology and ecology. These data are useful for both quantitative modelling and in providing background information about an area. The data have been collected from a large number of sources – survey data in agricultural economics, ecology and soils, measured data in meterology and land topography. Much of the information is at different scales, with a range of associated errors, and is, in many cases, incomplete. Data which are to be used in NELUP must be checked and may need preprocessing into standardized file formats.

The database is the core around which all information flows in NELUP (Figure 1.5) are organized. Various transformations and analyses are carried out on the database in order to derive composite datasets, to distribute data spatially, check errors and develop models. Most of the data manipulations are carried out using standard GIS software, but some are derived from algorithms and decision rules specially developed in the course of the project. Some data files used in the DSS are preprocessed from the database by a series of 'loading' routines in order to speed up the operation of the interactive parts of the system.

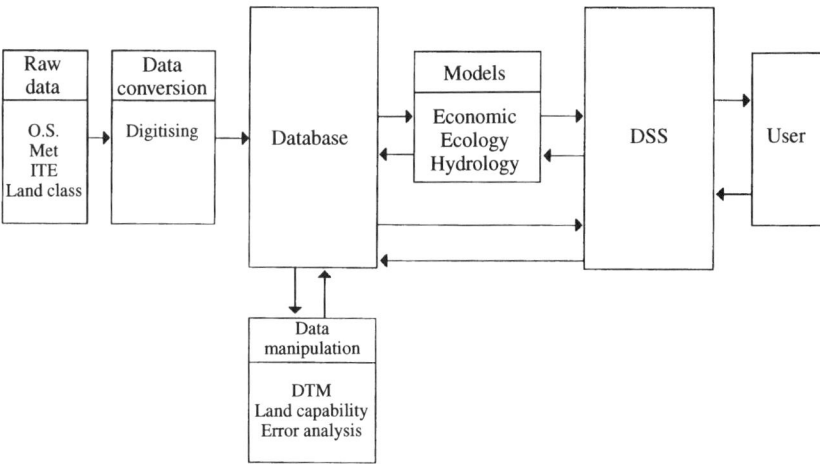

Figure 1.5 Data flows within the NELUP system.

Variables for running the models are drawn from the database. The models can be run in real time where run-times are short enough to be compatible with interactive operation. In the case of long runs, either the results for representative sets of conditions can be calculated and stored for future use, or the user can specify inputs to the model for processing later by the computer. The time to run the ecological model is measured in seconds, the economic and macro hydrological models in minutes, and the SHE may take several hours.

1.4.5 THE SCREEN DISPLAY

Graphical display at a work station terminal is the principal mode of communication between the user and the NELUP system. The guiding principle in the design of the display, which has been developed especially by NELUP, is that it should be user-friendly. The display commands are easy to understand and quick to execute, and the system encourages the user to interact with it. The system may be used both as a comprehensive source of background data on land use that can inform decision making, and as a simulator with the capability for comparing policy options at various levels of detail.

Printed words are not really the most appropriate way to convey information on spatially disaggregated problems. It is much more effective to distribute the information related directly to its location on the ground. By the nature of the subject, most of the information about land use is best conveyed on maps, but tables, histograms and

x–y plots may be used where they are appropriate. Computer graphics are ideally suitable for conveying information in a self-explanatory form that is clear, consistent and uncluttered. When hard copy is required, the output on the screen can be transferred to a printer.

1.4.6 THE DECISION SUPPORT SYSTEM (CHAPTERS 5–8)

Decision Support Systems are defined as computer-based information systems that combine models and data in an attempt to solve poorly structured problems with extensive user involvement.

Almost every decision about land use involves several interests, some of them conflicting. It will be influenced by the policy environment within which the decision should fit and it must conform with existing and proposed legislation. It must be economically viable and socially acceptable. It will most likely be influenced by technological change and, as agriculture is such a predominant user of rural land, the decision is often likely to be dependent on what takes place in agriculture.

Three pathways, and three sets of models, have been chosen for organizing the information relevant to decisions about land use and for simulating the processes involved as well as the linkages between them. The economic model interprets economic and social trends in order to predict changes in land-use policy and evaluate the likely economic outcomes of a proposed decision. The hydrological unit can simulate quantitatively the surface water and ground-water flows that result from a particular use of land, as well as the quality of the water that is likely to result from chemical inputs at the land surface. The ecological unit predicts the assemblages of plant, vertebrate and invertebrate species that result from a particular land use, together with an indication of the long-term consequences.

The decision support system attempts a synthesis of the information contained in the database through the medium of the mathematical models as shown diagrammatically in Figure 1.6. The models are used to simulate how a system of land use functions at the river basin scale, in order to make the results available in a form that can assist decision makers in reaching strategic choices about land use. The system should support rather than replace expert judgement.

Inputs to the models may be in the form of 'controllable' variables such as commodity prices, quotas, environmental restrictions, which the decision maker can hope to influence, together with uncontrolled variables such as weather and soil type, which have to be treated as exogenous to the system. The output from the models are the computed outcomes of the different policy options as they affect, for example, the income of the landholders, the ecology and the hydrology. The model-

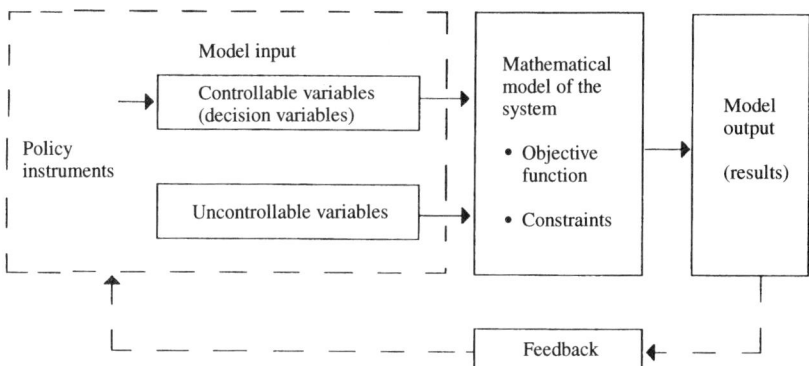

Figure 1.6 The decision support system.

ling capability of NELUP allows the policy options to be analysed at different levels of detail as required by the user. If the outcomes of a policy are deemed unsatisfactory, the decision maker can react by 'feeding back' a modified policy into the system for further analysis.

Although the NELUP methodology is of general application, it was developed using two river basins, the Tyne and the Cam, as examples. The Tyne, which contains 17 of the 25 land classes in the Remote-Sensed Classification of Land Cover, is typical of Upland Britain. The Cam is an example of an intensively managed lowland catchment. A river catchment is the natural unit of study for hydrology, where the prediction of the quantity and quality of the flow in the river is the principal objective. It also provides a useful check on the accuracy of the models, in that it allows the simulations to be compared with the historical gauging records for the rivers. However, the approach may be used to examine land use in any regional area, whose boundaries can be defined and for which reliable data are available. Remote-sensed, land-cover data are available at the pixel scale (25 m × 25 m) and can be used with a fair degree of confidence at an area of one hectare. In fact, land use studies should be approached at a range of scales, between the broad generality of the national scale at which the majority of policies are made, through 'natural areas' of similar biogeography to the different scales of particular habitats.

References

Department of the Environment (1990) *This Common Inheritance: Britain's Environmental Strategy*, Cm 1200, HMSO, London.

United Nations Development Commission (1993) *Agenda 21: Action Plan for the Next Century*, UNDC, New York.

2 The economic model

2.1 Introduction

Although agriculture and forestry employ a small proportion of the total workforce and contribute a relatively minor amount to national output, they are often of local economic importance and remain the dominant land use in a majority of river catchments within the United Kingdom. This means that catchment management plans concerned with, for example, water quality or habitat protection cannot be devised without due consideration of the factors influencing the production decisions of farmers and foresters and how such decisions may alter in response to changing market conditions and policy incentives or constraints. This chapter outlines briefly the analytical framework offered by neoclassical economics for addressing the determinants of agricultural and forestry land-use patterns and highlights some of the practical issues encountered in empirical economic modelling of land-use change within NELUP.

2.2 An analytical economic framework

The basic premise underpinning neoclassical economic theory is that economic agents are motivated by self-interest. In the case of producers (as opposed to consumers), this translates conventionally into an assumption of profit-maximizing behaviour. That is, private production decisions, including the allocation of land between alternative uses, are taken with the objective that profits accruing to each individual producer concerned are maximized, subject to constraints imposed by prevailing technology, resources and policies. Although this assumption may appear to be extreme, it does offer a powerful analytical approach for exploring land use and land use change and has a long pedigree, stretching back to the nineteenth century works of von Thünen and Ricardo. Some alternatives to the assumption of profit-maximizing behaviour are discussed at a later point in this chapter.

Given profit-maximizing behaviour, the land use observed at a particular location will depend on the relative profitability of competing potential uses. This in turn will depend on the production possibilities facing the land user and the prevailing market and policy conditions. The full set of production possibilities is defined by what is technically feasible, but the producer often faces a restricted set due to individual resource constraints, for example a limited land area or a fixed labour supply, or policy requirements such as limits on livestock numbers or fertilizer applications. The technical, physical relationships between inputs and outputs are delineated by a production or transformation function. Inputs include items such as land, labour, machinery and fertilizer, whereas outputs include (marketable) products such as wheat, barley, meat, milk and timber. Outputs also include by-products such as nitrate pollution, habitat fragmentation and attractive landscapes. These latter types of output represent unintended effects of private land-use decisions and are referred to as externalities, that is their costs or benefits are external to, and do not therefore affect directly, the decisions of individual profit-maximizing producers. The existence of externalities arising from agriculture and forestry is one reason why economic analysis of land-use patterns is pivotal to integrated catchment management.

Due to the central role of land as a productive resource and as an environmental sink for agricultural and forestry, production functions facing individual producers reflect site-specific characteristics such as soil types, climatic conditions and topography. Given the spatial variability in such site characteristics, it should not be surprising that production functions vary spatially. This results in different production possibilities being potentially available to producers at different locations. To take a simplistic example, good quality land in a fertile river valley can support a wide range of intensive agricultural activities whereas poor quality land at the top of a hill can support a narrow range of extensive activities. The terms intensive and extensive refer to the ratio of inputs to land area, with intensive activities being characterized by a high usage of inputs, such as fertilizer or livestock, per hectare of land and extensive activities characterized by a low usage of inputs. The distinction between different intensity levels illustrates that land use represents a combination of land cover and the intensity at which that cover is managed. Thus, for example, wheat receiving 200 kg/ha of nitrogenous fertilizer is distinct from wheat receiving 150 kg/ha of nitrogenous fertilizer. The distinction is not merely semantic since differences in production intensity translate into differences in the quantity of physical outputs per hectare, including externalities.

Although spatial heterogeneity of production functions may contribute to observed variation in land-use patterns, other factors are also important. In particular, individual producers typically face different resource constraints arising from differences in, for example, the physical or financial size of the business, current production patterns or mode of land ownership. Production possibilities may also be restricted directly by policy constraints, for example, bans on the use of certain chemicals or maximum livestock grazing densities. This means that individual producers may face different restricted production possibilities even if their physical site characteristics define the same full set of production possibilities. Consequently, profit-maximizing land-use mixes, that is the pattern of land covers and intensity of their management, would be expected to vary between producers. The precise land-use mix selected from the restricted production possibilities facing an individual producer will depend on the relative profitability of the available production possibilities, which in turn will depend on the prices of inputs and outputs associated with each production activity. Other things being equal, lower input prices and higher output prices will increase the profitability of a production activity, making it more attractive to a producer.

The above is a necessarily brief outline of the conventional neoclassical economic basis for addressing land-use change. Essentially, land-use patterns are determined by the private production decisions of individual farmers and foresters, each of whom chooses a profit-maximizing land-use mix from a set of restricted production possibilities given prevailing input and output prices. Changes in land use may arise therefore if the production possibilities available to a producer alter or if the relative profitability of available production possibilities alters. The latter will occur if input or output prices change, whereas the former may occur if planning or policy constraints tighten or relax, if technological advances change the production function(s), or if resource constraints faced by individual producers alter. Quantitative empirical implementation of this analytical framework facilitates exploration of the land-use change consequences of alternative market and policy conditions. In particular, it highlights spatial and temporal substitution effects both between separate land covers and between different intensity levels of a single land cover. The next section discusses empirical implementation of the framework within NELUP with respect to the catchments of the Tyne and Cam rivers.

2.3 Choice of empirical modelling technique

The neoclassical approach to production economics can be applied empirically through a variety of quantitative modelling techniques (Just, 1993). However, within NELUP, choice of empirical modelling technique was constrained by several requirements factors. First, the role of the economics group within NELUP is to forecast patterns of land use and associated resource use arising under different market and policy scenarios specified by users of the DSS. This means that the economic model must be sufficiently flexible to cope with a wide range of conditions, many of which may well extend beyond the realm of historically observed situations. Unless it is assumed that production decisions observed in the past will be unaffected by novel market and policy conditions, modelling techniques which rely solely on extrapolation of historical relationships would clearly be inappropriate.

Second, if the environmental impacts of production decisions induced by changing market and policy conditions are to be addressed, economic activity has to be expressed in physical as well as financial units. For example, ecological and hydrological impacts will vary with the number of livestock grazing a parcel of land rather than the value of those livestock. This means that economic modelling techniques which neglect the physical basis of production would be inappropriate.

Third, since the environmental impacts of land use depend on the spatial configuration of land uses as much as on the nature of land use at a particular location, the economic model has to include an explicitly spatial dimension. For example, aspatial predictions such as 'the area of arable land in the catchment as a whole increases by 200 ha' are inadequate. This means that economic modelling techniques which are not well suited to incorporating a reasonable (i.e. subcatchment) degree of spatial precision would be inappropriate.

Fourth, due partly to the three constraints mentioned already, the structure of the economic model must permit easy transfer of data to and from the DSS. Modelled policy decision variables, such as prices and regulatory controls, must be amenable to change from within the DSS, and output data must be generated in a consistent and manageable form. The model must also run sufficiently quickly for the DSS to operate on-line. This suggests that overly complex models may be inappropriate.

Fifth, as with the ecological and hydrological models, the economic model has to be based as far as possible on publicly available data and accepted theoretical principles, with minimum recourse to unique local surveys and *ad hoc* modelling techniques. This means that innovative modelling techniques with high data demands are unlikely to be tenable.

Although various modelling approaches were considered, the above constraints effectively dictated the choice of mathematical programming techniques, specifically linear programming, for empirical economic modelling within NELUP. This approach essentially involves defining a set of production activities, in terms of input–output relationships, and searching across this set to find a combination that optimizes an objective function subject to constraints imposed by, for example, resource availability or policy restrictions. Formally, a linear programming model may be expressed as:

Maximize (or minimize): $$Z = \mathbf{cx} \quad (2.1)$$
Subject to:
Linear production processes: $$\mathbf{Ax} \leq \mathbf{b} \quad (2.2)$$

The objective function (2.1) simply multiplies a vector of outputs (**x**) by a vector of output prices (**c**). The expression (2.2) represents constraints on production, although additional constraint lines can be added. The matrix **A** consists of input–output coefficients describing the technical transformation of inputs into outputs, with **Ax** defining the level of production, which is constrained to use no more than the available resources, **b**. Additional constraints can easily be imposed on the level of particular elements of the **x** or **b** vectors.

Linear programming (LP) satisfies the empirical economic land-use modelling needs within NELUP in several respects. First, programming models can be constructed using minimal datasets derived from a variety of sources: programming models are relatively unconstrained by the availability of historical data and can be used to forecast production patterns 'out-of-sample'. This is in sharp contrast to econometric techniques which are often favoured in forecasting exercises, because correctly specified models can extract the maximum amount of information from sample data and give an indication of the statistical reliability of results, but are typically unsuited to projecting situations where production conditions are likely to stray outside of the historically observed sample on which the model was estimated (Shumway and Chang, 1977).

Second, the structure of programming models allows a wide range of price and non-price policy instruments to be specified. This is undoubtedly useful given the emergence of input and output quotas and cross-compliance measures in agricultural and environmental policies. In particular, the ability to impose equality or inequality constraints on individual elements of the **x** and **b** vectors mimics policy measures such as limits on livestock grazing densities or land set-aside designations.

Third, the representation of production relationships by input–output coefficients lends itself equally to financial or physical measures of economic activity. The **A** matrix is simply a set of input–output coefficients describing the transformation of inputs into outputs. As such, elements of **A** may be expressed in financial terms or physical terms. In the latter case, input–output coefficients may include items such as nitrate pollutants or habitat degradation arising from a particular land use.

Fourth, by designating blocks of (not necessarily contiguous) land with identifiably different production possibilities, it is relatively easy to incorporate a spatial dimension within LP models. Although this may not account for the precise location or shape of individual parcels of land, using land blocks or classes offers a convenient means of moulding an aspatial modelling technique to address spatial problems (Chuvieco, 1993). In particular, the flexibility of specification of elements of the **A** matrix allows input–output coefficients to include spatial relationships between land uses on different parcels of land, for example the fact that upland lambs are reared on lowland farms or that irrigation usage upstream affects production possibilities downstream.

Fifth, LP models can run recursively over several consecutive time-periods with the model results of one time-period serving as the starting conditions for the next time-period (Day, 1963a). This is useful given the time-lags inherent in agricultural and forestry production: the biological and natural resource base of these land uses ensures that change is not instantaneous. For example, livestock numbers take time to increase naturally, grassland does not improve or degrade overnight and land users rarely escape resource constraints spontaneously. The fourth and fifth points are crucially important if the economic model is to be capable of identifying knock-on effects of changes in one land use on other land uses.

The above characteristics of linear programming perhaps account for its popularity in applied agricultural economic studies, particularly where only minimal datasets are available (Taylor and Howitt, 1993). Consequently, procedures for constructing LP models are reasonably well established and, once a base-line framework has been designed for the problem in hand, it can be used to tailor a model to fit different locations. In addition, LP models can be constructed and run using a variety of software packages, including SAS (SAS Institute, 1991), a powerful statistical package that runs under UNIX and is able to transfer data to and from the DSS. Hence the use of LP modelling satisfies all of the constraints identified above. More detailed explanations of linear programming are offered by Dykstra (1983) and Hazell and Norton (1986). The components of the NELUP economic LP models are shown in Figure 2.1.

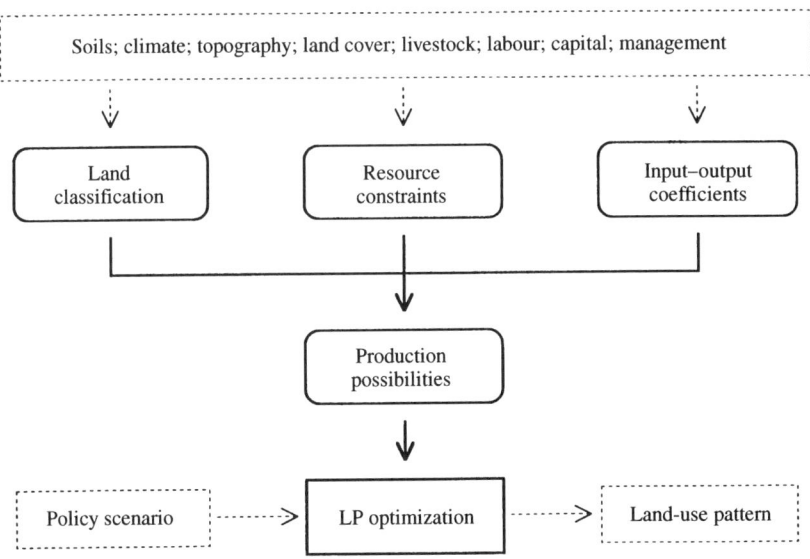

Figure 2.1 Components of NELUP economic LP models.

2.4 Data collation

The data collation procedure for the economic modelling component of NELUP essentially fell into three stages, each of which were followed for the two study catchments. The first stage involved using a geographical information system (GIS) to combine various digital environmental datasets in order to classify the catchment land resource base into different land classes. For the purposes of NELUP, the agricultural land capability classification devised by the Macaulay Land Use Research Institute (MLURI) was used (Bibby *et al.*, 1991). This classification system incorporates a number of factors of environmental significance, such as soil type, climate and topography, and therefore offers a convenient common spatial backcloth for the three modelling groups of economics, ecology and hydrology. Other land classifications may be more appropriate for different modelling tasks.

The second stage of data collation involved identification of production possibilities for each land class. In principle, an infinite number of technically feasible activities could be imagined on each land class. However, modelling an infinite range of activities is clearly intractable and a smaller subset has to be used. There is a trade-off here in that the smaller the subset, the easier it is to model but the greater the risk of imposing preordained limits on the mix of land uses that will be forecast: an LP model can only pick from the range of activities it is offered.

For the purposes of NELUP, production possibilities were based on historical land uses reported in the agricultural census (Moxey and Allanson, 1994), commonly observed land uses on different MLURI capability land classes (Bibby *et al.*, 1991), commonly observed land uses on different soil associations in the study areas (Hodge *et al.*, 1984; Jarvis *et al.*, 1984), observed production activities reported by the Farm Business Survey (FBS), biophysical constraints on biological production systems in the study areas, and expectations of farmers and other experts about likely future agricultural land uses. Attention was focused on land-based agricultural enterprises that are, or potentially could be, profitable for a large number of farmers and therefore currently, or potentially, occupy large areas of land across the catchments. Activities occupying little land, for example intensive poultry rearing, and minor, more specialized activities that are restricted to a few farms due to limited market opportunities or biophysical constraints were disregarded. Although this may neglect some locally important activities, it is compatible with the aims of a DSS designed to support decisions at the regional rather than local level.

The third stage of data collation involved estimating input–output coefficients for production possibilities on each land class. Each production activity has several coefficients associated with it, reflecting both the quantity of inputs required and links to other activities. Coefficients for the **A** matrix may be taken from a variety of sources including published 'industry standards', agronomic field trials and (informed) 'expert opinion'. However, these may be inappropriate for developing a catchment model since potentially important regional differences in production characteristics may be masked. For this reason, advantage was taken of access to FBS data which offers the opportunity to derive linear production coefficients from systematically observed farm behaviour. Although the conventional farm accounting procedures employed in the FBS mask specific input–output relationships by reporting total input usage across a farm rather than per enterprise, and only input expenditure, not physical quantities, is recorded, following Tyler (1966) and Errington (1989), it is possible to apportion total input usage between individual enterprises by using a 'form of derived demand' model:

$$\mathbf{b} = \mathbf{Ax} + \mathbf{e} \qquad (2.3)$$

where $\mathbf{b}$ and $\mathbf{x}$ are, respectively, input expenditure and output revenue vectors, $\mathbf{A}$ is a matrix of input–output (I–O) coefficients and e is a vector of random disturbances. Individual elements of the $\mathbf{A}$ matrix thus represent the average amount of expenditure required on a particular input to produce a £'s worth of a particular output.

Errington (1989) shows that using ordinary least squares (OLS) leads to negative production coefficients where, *a priori*, all such coefficients are assumed to be non-negative. Although this problem may be tackled using a variety of techniques to impose non-negativity within a 'Classical' statistical approach (Ray, 1985), the approach adopted within NELUP was to use a Bayesian framework (Griffiths, 1988). Essentially this involves using Bayesian priors to combine the assumption of non-negativity with the sample information embodied in the OLS estimates (Chalfant *et al.*, 1991; Moxey and Tiffin, 1994).

Due to the reporting base for FBS data, I–O coefficients estimated in this manner represent average production relationships. More detailed nitrogen and irrigation response coefficients were estimated using the Erosion Productivity Impact Calculator (EPIC), a biophysical crop-growth simulation model developed by the United States Department of Agriculture (USDA) in the early 1980s. The use of crop-simulation models overcomes the lack of experimental data and is increasingly commonplace in economic studies (Bouzhar *et al.*, 1990; Johnson *et al.*, 1991), representing an approximation of the engineering production function concept proposed by Chenery (1949).

Crop yields (including grass) under alternative fertilizer, irrigation and grazing management regimes were simulated using EPIC, and run for climate–soil combinations determined by the GIS. Weighted average yields were then calculated for each crop within each MLURI land class. These estimates enhance the spatial detail of the LP model by accounting for yield variability across different land classes, and also enable differences in management intensity levels to be incorporated into the LP model. EPIC is also used within the NELUP hydrology model.

Coefficients derived from the FBS data and EPIC simulations were supplemented by information taken from the literature. In particular, information on the relative grazing value of different grasslands, costs of improving grassland and forestry practices were derived from reported agronomic research and recommendations (for example, SAC, various years).

2.5 Model construction

Once data collation was complete, the various data items were assembled within a formal LP optimization framework. In the case of the two pilot study areas chosen for analysis by NELUP, each encompasses over 1800 farm and forestry production units. In principle, catchment-level land-use predictions could be achieved by modelling each individual farm. In practice, the data necessary for this are not available. Moreover, using 1800 separate LP models would preclude the DSS from running on-line.

Catchment-level results could be obtained by running a set of representative farm models and multiplying results according to the frequency of each farm type in the catchment. Unfortunately, available data on the joint size and type distribution of farms are not sufficient to permit this (Day, 1963b). It is also difficult to incorporate inter-farm links, such as lambs born on upland farms being sold to lowland farms for fattening, into a set of representative farm models. This may be important since inter-farm links determine, at least partially, knock-on effects of land-use change in one part of the catchment on other parts of the catchment. Consequently, although farm-level models provide useful insights into farm-level responses and patterns of land-use, they are not particularly useful for catchment-wide predictions.

For this reason, catchment-scale land-use modelling within NELUP is achieved through an aggregate LP model which treats the catchment as a single farm. The purpose of the catchment-level model is to integrate the information on financial and physical production processes in a form which accounts for the distribution of changes in the pattern and intensity of crop and livestock production across different land classes.

The use of an aggregate-level LP model is a widely accepted means of modelling large areas (Norton and Schiefer, 1980). However, a catchment scale model of land use will overstate flexibility and co-ordination of production. Essentially, bias arises because a study area typically encompasses a number of non-identical production units (farms), yet, although information may be available on the total resource base, information on the structure of each individual production unit is typically incomplete. A simple example will serve to illustrate this.

Consider a catchment with two land classes and two farms, each wholly on separate land classes. The farms grow two crops (x) which receive different profit rates (c) reflecting different cost and yield levels on the land classes. Each farm grows the crops using just two resources, land and capital (b), which are available in fixed quantities. The technical relationships between these resources and the crops are given by input–output coefficients (a). The LP form of this problem for each farm is:

Farm 1

Max $c_{110}x_{110} + c_{210}x_{210}$

subject to

$a_{111}x_{110} + a_{211}x_{210} \leq b_{011}$
$a_{112}x_{110} + a_{212}x_{210} \leq b_{012}$

Farm 2

Max $c_{120}x_{120} + c_{220}x_{220}$

subject to

$a_{121}x_{120} + a_{221}x_{220} \leq b_{021}$
$a_{122}x_{120} + a_{222}x_{220} \leq b_{022}$

where each element in the LP formulation has three subscripts, i, j, k. The first subscript denotes the crop (1,2), the second denotes the land class (1,2) and the third denotes the resource (1 for land, 2 for capital). A '0' subscript indicates that the particular distinction is not relevant for that element.

If only the total amount of capital used in the region is known ($b_{012} + b_{022}$), it would be necessary to run the model as an aggregate LP of the form:

Aggregate model

Max $c_{110}x_{110} + c_{210}x_{210} + c_{120}x_{120} + c_{220}x_{220}$

st

$a_{111}x_{110} + a_{211}x_{210} \leq b_{011}$
$a_{121}x_{120} + a_{221}x_{220} \leq b_{021}$
$a_{112}x_{110} + a_{212}x_{210} + a_{122}x_{120} + a_{222}x_{220} \leq (b_{012} + b_{022})$

Implicitly, this leads to relaxation of the capital constraint and can only lead to an increase in total profits, thus the aggregate model tends to overstate the degree of resource flexibility.

Table 2.1 Example of LP solutions

Crop	Aggregate model	Summed farms
x_{11}	0.0	0.0
x_{21}	100.0	100.0
x_{12}	16.7	66.7
x_{22}	133.3	83.3
Profit	32 500.0	30 200.0

For instance, for the specific example where:

$c_{110} = 100$, $c_{210} = 150$, $c_{120} = 90$, $c_{220} = 120$,

$b_{011} = 100$, $b_{021} = 150$, $b_{012} = 90$, $b_{022} = 110$,

$a_{111} = a_{211} = a_{121} = a_{221} = 1$, $a_{112} = 0.5$, $a_{212} = 0.7$, $a_{122} = 0.6$, $a_{222} = 0.9$

If the problem is formulated as an aggregate LP model with the capital constraint specified as $(b_{012} + b_{022})$, then profit is estimated to be 32 500. However, if the problem is formulated as two separate LPs, then profit is reduced to 30 200. These results are presented in Table 2.1.

The problem of aggregation bias would be resolved if there was information on capital resource available on both farms instead of the aggregate level. The problem in this case is worsened by the existence of different types of land resource on each of the farms.

Thus, although a catchment-level LP model does offer a convenient means of modelling land use at the catchment scale, it may incur aggregation bias. For this reason, a set of representative farm models was also developed within NELUP to provide insights into land-use change at a more local level (Oglethorpe and O'Callaghan, 1995). Both sets of models shared common data but, in the absence of data on the joint size and type distribution of farms, no attempt was made to reconcile farm-level model results with catchment-level model results.

2.6 Validation

If the DSS is to provide effective support to potential users, the models that underlie it must be demonstrably reliable. This means that the economic LP models had to be validated and limits on their predictive performance identified. Unfortunately, rigorous validation is hindered by a lack of suitable benchmark data. In particular, data on past land use is incomplete: not only do agricultural census and satellite data describe only land cover, not land management, but census data have not been released at the parish level since 1988 and available satellite data relate only to 1990. Nevertheless, some validation was possible, albeit not as comprehensive as might be hoped (McCarl, 1984; Hazell and Norton, 1986). First, individual I–O elements of the LP models were subject to scrutiny. Second, actual model performance was explored.

Due to the incorporation of different land classes, each characterized by a range of production activities, and links between activities on different land classes plus the need to tie activities in one year to activities in another year, the 'A' matrix of I–O coefficients is rather large, with several thousand non-zero elements. In principle, each of these should be validated individually. In practice, this is not possible since the number of coefficients is simply too large and, perhaps more importantly, there are few 'hard' data against which to compare them. Nevertheless, selected I–O coefficients were investigated.

Coefficients estimated from FBS data were compared with other published sources and shown to local experts. In general, they were deemed to be plausible, but it should be noted that in some cases they may not match estimates derived from other sources due to differences in the basis of estimation or reporting (Moxey and Tiffin, 1994).

Validation of the use of EPIC to estimate crop yields was also hampered by a lack of 'hard' data. Due to a lack of detailed yield and management data, direct comparison of simulated and observed yields was restricted to field trial data at a few individual experimental (University) farms within the study areas. However, some comparison with yields recorded in the FBS was also possible. In both cases, simulated yields were within $\pm 11\%$ of observed, a generally acceptable level of precision for biophysical simulation (Dent and Blackie, 1979). In addition, although available data are insufficient to verify that the magnitude of yields estimated for different MLURI land classes and different nitrogen fertilizer and irrigation water input levels, the ranking of yields was judged reasonable by local experts. In addition, EPIC has been used extensively world-wide, and has been found to perform well under a wide range of conditions in various countries (Engelke and Fabrewitz, 1991; Favis-Mortlock et al., 1991; Jones et al., 1991; Flichman et al., 1994).

Validation of actual model performance took the form of running the LP models under observed historical market and policy conditions, and comparing model results with actual observed economic activity during the 1980s. In the case of the catchment-level models, this meant comparing results with land-use areas and livestock numbers reported in the agricultural census. In the case of the farm-level models, model results were compared with more detailed information provided for individual farms within the Farm Business Survey. In both cases, the LP models were found to perform adequately. However, some errors were present.

In particular, the catchment-level models persistently underestimated the area of some land uses whilst overestimating others. Indeed, a summary measure of this error – the percentage absolute deviation (PAD)* – reveals an error of between 10% and 15%. This apparent model inadequacy arises for several reasons. First, there is uncertainty over the robustness of some of the benchmark land-use data used for comparison with model results. For example, the distinction between different types of grassland is somewhat blurred. Second, bias due to aggregation and the deliberate exclusion of some minor land uses is undoubtably present. Third, the representation of production possibilities by linear technology tied to specific land classes may be too restrictive in some circumstances. However, although the magnitude of the PAD errors may appear large, they are within the bounds of errors reported for other aggregate-level LP models (Norton and Schiefer, 1980). Moreover, although an alternative form of model (i.e. an econometric formulation) might track historical patterns more closely, LP models still retain the advantage of modelling 'out-of-sample'.

2.7 Model usage

The economic models are incorporated within the NELUP DSS. As such, they are available to users of the DSS wishing to explore the consequences of alternative policy scenarios. The DSS interface to the models allows users to alter a range of input and output prices, to specify input and output quotas and to designate areas where explicit production constraints are imposed. The LP models take the specified scenario details and generate predictions of, for example, profit levels, areas of different land uses, employment levels, numbers of livestock, irrigation levels and usage of fertilizers at either the farm or catchment level. This information is displayed directly through the DSS but is also routed internally to the ecological and hydrological models. Specimen example case studies using the economic models are reported elsewhere in this book. For illustrative purposes, Tables 2.2 and 2.3 are also included here to show the type of summary output obtained from the LP models. The results refer to two policy scenarios applied through the catchment-scale LP model for the Tyne catchment. The first scenario gives a baseline relating to the current agricultural situation. The second scenario gives a forecast of the situation following imposition of a catchment-level quota for inorganic nitrogenous fertilizer 10% below baseline usage – a policy which might be pursued as a means of reducing nitrate pollution (Moxey and White, 1994).

The results show that imposition of the nitrogenous fertilizer quota leads to a reduction in profits through a dramatic reduction in revenue,

---

* This is defined as the average absolute deviation between the predicted and the observed areas divided by the average actual value, and is a commonly used error measure in LP modelling (Norton and Schiefer, 1980).

Table 2.2 Financial land use performance (£ 000s, 1988 prices)

Scenario	1: Baseline	2: N quota
Total revenue	42 432	33 432
Total costs	−38 953	−30 683
Profit margin	3 479	2 749

Table 2.3 Physical land-use performance

Scenario	1: Baseline	2: N quota
Arable (ha)	20 500	15 625
Temporary grass (ha)	11 866	13 020
Permanent pasture (ha)	66 125	55 455
Rough grazing (ha)	108 739	123 130
Total N use (t)	18 500	16 700
Total livestock units	101 642	99 300

although this is offset to some extent by a reduction in costs. The reduction in revenue reflects primarily a decline in the area of arable crops as marginal land is switched from cereals to grassland. In turn, some poorer quality pasture degrades to rough grazing as fertilizer applications are withdrawn. The reduction in costs is attributable to lower fertilizer expenditure, but also to savings on labour and machinery associated with the displaced arable production. More detailed tabular output can be extracted from the LP models, but the results presented above are a good summary of the type of information that can be generated.

2.8 Some alternatives to profit maximization as an objective function

The assumption of constrained profit-maximizing producer behaviour underpins the empirical economic models discussed above and is reflected in the objective functions specified. However, it is possible to envisage alternative objective functions. Two will be considered here: incorporation of producers' attitudes to risk; and framing the objective function in terms of a policy body's objectives rather than producers' objectives.

Although profit-maximizing behaviour provides a simple rule-of-thumb for modelling production decisions, it neglects the possibility that producers may respond to profit variability as well as mean (expected) profit levels. Given the biological nature of agricultural and forestry production, profits may be subject to some uncertainty due to stochastic natural factors plus variable input and output prices. If some producers opt for land-use patterns giving a lower absolute level of profit but less variability, the land-use consequences could be significant. This issue has been explored to a limited extent with the farm-level models (Oglethorpe and O'Callaghan, 1995). Unfortunately, a switch away from the profit-maximizing assumption to a more sophisticated representation of producer behaviour is impractical for large-scale studies. The empirical evidence regarding attitudes to risk is often inconclusive (Bond and Wonder, 1980) and there are dangers inherent in assuming some average or representative attitude to risk for all producers within a large area. Further, the information base required to estimate such a 'representative' attitude to risk is often prohibitively expensive to collect since detailed surveys of individual producers are required. Moreover, inclusion of risk-attitudes within an LP model increases the size of the model dramatically since each production activity can no longer be represented by a single (mean) I–O coefficient. For these reasons, the assumption of profit maximization was retained for LP models within the DSS.

A second alternative to the profit-maximizing assumption embodied in the catchment-level LP model would be to replace the producers' profit-maximizing objective function with an objective function reflecting the goals of a policy body. As currently framed, the catchment-level LP predicts land-use patterns arising under specified market and policy conditions as if the catchment operates as a single profit-maximizing farm. A DSS user wishing to achieve certain goals, for example a target level of water quality or a minimum level of rural employment, can use the DSS to explore 'what if' scenarios to see if the goals are achievable. This reflects the premise that policy bodies cannot change land-use patterns directly but rather have to operate indirectly by restricting the options available to land users or altering relative profitability. However, this 'trial and error' procedure may be time-consuming for users of the DSS.

An alternative would be to reframe the objective function in terms of goals of a policy body and to assume that policies could be imposed unilaterally. Such an approach is referred to as multicriteria programming (MCP) or multicriteria decision making (MCDM) and would allow more direct imposition of desired goals (Zeleny, 1982; Romero and Rehman, 1989). Given the structure of the catchment LP, there may be some potential for using the LP model as a MCP model to determine the constrained 'socially optimal' production pattern within the catchment as a way of indicating which policies, in terms of taxes or quotas, are appropriate. Essentially this would involve identifying the constraints in the LP model which are to be treated as goals, and establishing the required weights or priority levels attached to each goal by the policy body concerned. However, there is a potential danger here in that the LP model was designed to reflect the behaviour of a large number of profit-maximizing individuals and may not be immediately suitable for conversion to a MCP model: the individual goals of private land users will only coincide with the aggregate goals of the planner if the incentives are in place to bring about the required modification of production patterns. For this reason, MCP modelling has not yet been addressed within NELUP.

2.9 Conclusions

The role of the economics group within NELUP is to model the response of agricultural and forestry land use to changing market and policy conditions. Such modelling is of relevance to catchment management since agriculture and forestry remain dominant land uses in many UK river catchments and the private decisions of individual land users can have significant social and environmental implications. Indeed, economic modelling of land use is concerned not only with the direct effects on agricultural inputs and outputs plus income and employment levels, but also the indirect or externality effects which lead to non-point water pollution and ecological change. Moreover, economic models are well suited to highlighting temporal and spatial substitution and knock-on effects induced by changes in prevailing market and policy conditions. For example, restricting land uses at one location displaces resources that may be redeployed elsewhere, with subsequent impacts.

The basic premise underpinning the NELUP economic models is that private land-use decisions are motivated by the pursuit of profit maximization, constrained by available technology, resources and policy restrictions. This behavioural model is implemented empirically within a linear programming framework, both at the catchment and farm level. Choice of modelling approach was largely dictated by the research brief and the general paucity of data relating to current land uses and production relationships. The particular task faced by the economics group concerns linking the private decisions taken at the level of the firm, in terms of what to produce and how to produce it, to a set of spatially and temporally distributed physical parameters which are of hydrological and ecological significance. In addition, the economic model has to interact on-line with a DSS and rely upon publicly available data.

Although the empirical models embody some strong assumptions and undoubtably incur some prediction errors, they have been shown to perform adequately. Although alternative modelling techniques could be employed, in the absence of comprehensive land-use data against which to compare model performances, the choice of modelling approach is to an extent arbitrary since a complete validation of competing models is not possible (Taylor and Howitt, 1993).

References

Bibby, J.S., Douglas, H.A., Thomasson, A.J. and Robertson, J.S. (1991) *Land Capability Classification for Agriculture*. Macaulay Land Use Research Institute, Aberdeen.

Bond, G. and Wonder, B. (1980) Risk attitudes amongst Australian farmers. *Australian Journal of Agricultural Economics*, 24, 16–34.

Bouzhar, A., Braden, J.B. and Johnson, G.V. (1990) A dynamic programming approach to a class of nonpoint source pollution problems. *Management Science*, 36, 1–15.

Chalfant, J.A., Gray, R.S. and White, K.J. (1991) Evaluating prior beliefs in a demand system: the case of meat demand in Canada. *American Journal of Agricultural Economics*, 73, 476–90.

Chenery, H. (1949) Engineering production functions. *Quarterly Journal of Economics*, 59, 507–31.

Chuvieco, E. (1993) Integration of linear programming and GIS for land-use modelling. *International Journal of Geographical Information Systems*, 7, 71–83.

Day, R.H. (1963a) *Recursive Programming and Production Response*, North-Holland, Amsterdam.

Day, R.H. (1963b) On aggregating linear programming models of production. *Journal of Farm Economics*, 45, 797–813.

Dent, J.B. and Blackie, M.J. (1979) *Systems Simulation in Agriculture*, Applied Science Publications, London.

Dykstra, D.P. (1983) *Mathematical Programming for Natural Resource Management*, McGraw Hill, New York.

Engelke, R. and Fabrewitz, S. (1991) *Simulation runs with the EPIC model for different data sets*. Unpublished manuscript, Arbeitsgruppe System-forschung, Universitat Osnabruck, Germany.

Errington, A. (1989) Estimating input–output coefficients from regional farm data, *Journal of Agricultural Economics*, 40, 52–6.

Favis-Mortlock, D.T., Evans, R., Boardman, J. and Harris, T.M. (1991) Climate change, winter wheat and soil erosion on the English South Downs. *Agricultural Systems*, 37, 415–34.

Flichman, G., Garrido, A. and Ortega, C.V. (1994) *Agricultural Policy and Technological Choice: A Regional Analysis of Income Variation, Soil Use and Environmental Effects Under Uncertainty and Market Imperfections*. Paper presented to 34th EAAE Seminar, 'Environmental and Land Use Issues in the Mediterranean Basin: An Economic Perspective', Zaragoza, Spain 7–9 February, 1994.

Griffiths, W.E. (1988) Bayesian econometrics and how to get rid of those wrong signs. *Review of Marketing and Agricultural Economics*, 56, 36–56.

Hazell, P. and Norton, R. (1986) *Mathematical Programming for Economic Analysis in Agriculture*. Macmillan, New York.

Hodge, C.A.H., Burton, R.G.O., Corbett, W.M. et al. (1984) *Soils and Their Uses in Eastern England*. Soil Survey of England and Wales Bulletin 13, Harpenden.

Jarvis, R.A., Bendelow, V.C., Bradley, R.I. et al. (1984) *Soils and Their Uses in Northern England*. Soil Survey of England and Wales Bulletin 10, Harpenden.

Johnson, S.L., Adams, R.M. and Gregory, M.P. (1991) The on farm costs of reducing groundwater pollution. *American Journal of Agricultural Economics*, **73**, 1063–73.

Jones, C.A., Dyke, P.T., Williams, J.R. *et al.* (1991) EPIC: an operational model for the evaluation of agricultural sustainability. *Agricultural Systems*, **37**, 341–50.

Just, R.E. (1993) Discovering production and supply relationships: present status and future opportunities. *Review of Marketing and Agricultural Economics*, **61**, 11–40.

McCarl, B. (1984) Model validation: an overview with some emphasis on risk models. *Review of Marketing and Agricultural Economics*, **52**, 153–73.

Moxey, A. and Allanson, P. (1994) Areal interpolation of spatially extensive variables: a comparison of alternative techniques. *International Journal of Geographical Information Systems*, **8**, 479–87.

Moxey, A. and Tiffin, J.R. (1994) Estimating linear production coefficients from farm business survey data: a note. *Journal of Agricultural Economics*, **45**, 381–5.

Moxey, A. and White, B. (1994) Efficient compliance with agricultural nitrate pollution standards. *Journal of Agricultural Economics*, **45**, 27–37.

Norton, R.D. and Schiefer, G.W. (1980) Agricultural sector programming models: a review. *European Review of Agricultural Economics*, **7**, 229–64.

Oglethorpe, D. and O'Callaghan, J.R. (1995) Farm level economic modelling within the river catchment DSS. *Journal of Environmental Planning and Management*, **38**, 93–106.

Ray, S.C. (1985) Methods of estimating the input coefficients for linear programming models. *American Journal of Agricultural Economics*, **68**, 662–5.

Romero, C. and Rehman, T. (1989) *Multiple Criteria Analysis for Agricultural Decision Making*. Elsevier, Amsterdam.

SAC (1987–1990) *Farm Management Handbook*, various editions, Scottish Agricultural Colleges, Edinburgh.

SAS Institute (1991) *SAS/OR User's Guide, Version 6*, first edition, SAS Institute Inc., Cary, NC, USA.

Shumway, C.R. and Chang, A.A. (1977) Linear programming versus positively estimated supply functions: an empirical and methodological critique. *American Journal of Agricultural Economics*, **59**, 344–57.

Taylor, C.R. and Howitt, R. (1993) Aggregate evaluation concepts and models, in *Agricultural and Environmental Resource Economics* (eds G.A. Carlson, D. Zilberman and J.A. Miranowski), Oxford University Press, Oxford.

Tyler, G. (1966) The use of multiple-regression analysis of whole-farm data on the estimation of enterprise labour coefficients. *Farm Economist*, **11**, 106–19.

Zeleny, M. (1982) *Multiple Criteria Decision Making*, McGraw-Hill, New York.

3 Hydrological modelling

3.1 Introduction

Existing land- and water-use patterns are the result of many integrated actions which represent a dynamic and hydrologically complex background against which future water conditions must be assessed if the impacts of land-use change on the hydrology of a river basin are to be evaluated meaningfully. In order to establish the impact of future land-use change on the water resources of a river basin, it is necessary to understand the existing hydrological conditions of the basin and to quantify the extent to which the water resource will be modified. In addressing these two requirements a hydrological modelling system must account for the processes governing the movement and accumulation of water and contaminants throughout the basin and represent the controls on water movement and accumulation imposed by human actions. Furthermore, the system must provide descriptions of the state of the water in the river basin in forms that allow future events to be analysed and compared.

Illustrations of the potential problems to be analysed using a hydrological model are many and varied. For instance, alteration of stream and river flows may lead to increased flood risk as well as to the degradation of in-stream water quality and the development of unacceptable environments for aquatic flora and fauna. On land, soils can become waterlogged or conversely too dry to support the existing vegetation, affecting not only the distributions of plants but also the insects, invertebrates and mammals they support. Under some land-use scenarios, the quantity of water available for domestic, industrial and agricultural use can be reduced by increased losses to the atmosphere. Moreover, the quality of the resource can degrade below acceptable levels due to contamination from polluted effluent. Since any of these issues may be important under future conditions, the most appropriate modelling system is one in which all events/outcomes can be represented. Such a system must also be capable of predicting events for which no prior knowledge of the future state of the catchment is available. Finally, the system must be able to provide results efficiently for analysis alongside the results from the economic and ecological models.

Two modelling systems, NUARNO and SHETRAN have been implemented to provide the necessary hydrological modelling capability in the Decision Support System (DSS). The NUARNO modelling system provides an integrated description of catchment hydrological behaviour and is computationally efficient. It can be used to screen different future scenarios to determine the relative change in hydrological conditions. The SHETRAN modelling system provides a detailed spatial description of hydrological behaviour but is computationally relatively slow. It is used, therefore, to analyse the response of the worst-case scenarios predicted using NUARNO. Thus, the two systems are complementary.

The following sections introduce the two systems. In section 3.2 the processes represented in the modelling systems are described whereas in section 3.3 descriptions of the model components governing flow and transport are presented. The procedures for the use of the two systems are described in section 3.4. Finally, two specific issues relevant to the interpretation of hydrological predictions are presented in section 3.5, namely simulation scale and parameter uncertainty.

3.2 Hydrological processes

The processes governing the movement and accumulation of water and contaminants at any point within a river basin can best be identified by considering in turn the different paths taken by waters entering the river basin on their journey to the different points of exit. In this context, only liquid phase water movements are considered. Descriptions of the movement of atmospheric water vapour are normally neglected in hydrological models. Figure 3.1 shows the major paths for water movement in the land phase of the hydrological cycle for a rural river basin. The typical annual totals traversing each path are identified for the Cam river basin in the United Kingdom and serve to provide a guide to the major components of the water cycle for a heavily managed water resource system which includes a significant groundwater component.

The paths in the hydrological cycle illustrated in Figure 3.1 are numbered to assist the following brief description of the main processes in hydrology. Water enters a river basin through precipitation, as rain, hail or snow (1) and is intercepted at the land surface either by the vegetation or at the exposed soil surface. Precipitation onto the vegetation accumulates in the canopy (2) and the water eventually either runs off the plants to the underlying soil or evaporates to the atmosphere where it may recondense to provide further precipitation. At the soil surface, water infiltrates into the soil pores (3). If the intensity of rainfall exceeds the infiltration capacity of the soil then water will pond on the soil surface and may run off laterally within small channels (rills) or as sheet flow (4). The resulting overland flow infiltrates into the soil further

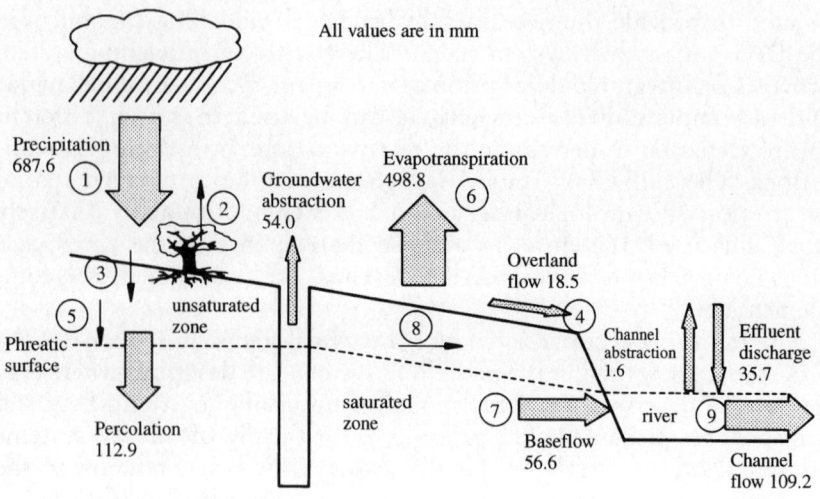

Figure 3.1 Major pathways in the hydrological cycle.

down the land slope or eventually runs into a water course. Water that enters the soil system may percolate downwards and accumulate in the soil profile (5) where it is accessible to plant roots or where it may be intercepted by lateral drainage systems such as tile drains. Water taken up by vegetation (6) is either retained within the body of the plants or is evaporated through the leaf structure of the plants to return to the atmosphere. Water retained in the soil can be drawn upwards to the soil surface by capillary forces where it may evaporate to the atmosphere. Water in the soil that bypasses the vegetation and drainage systems, and is not evaporated at the soil surface, will eventually pass down into the groundwater in the underlying rock strata where it can accumulate and migrate laterally (7). In layered soils, high infiltration rates can result in soil horizons near the land surface becoming saturated, leading to shallow lateral migration of water as interflow (8). Interflow may seep out downslope at the land surface or enter the stream and river systems. The stream and river channel network (9) provides the major natural drainage path out of the river basin, but water may also be lost by subsurface outflows through the aquifer system. Migration of water in stream channels is complicated by interactions with the adjacent aquifer system. Evaporation from open water surfaces takes place also. Where water accumulates in ponds, lakes and reservoirs, the amount of water removed by evaporation can be significant. The major natural outflows from the basin are therefore (a) evaporation to the atmosphere, (b) the outflow along the river network and (c) subsurface outflow through a groundwater pathway. At each point in the system, abstraction may take place for agricultural, urban and industrial use. Equally waste water can enter the water courses as industrial, urban and agricultural effluent.

Each of the pathways for water movement is also a pathway for contaminant migration and accumulation. Contaminants can be injected into any of the pathways within the hydrological cycle where they can move with the water, bind to the rock and soil surfaces or be taken up by vegetation. Contaminants may occur naturally from sources such as atmospheric deposition, soil chemical transformations or release from decaying vegetation. They can also be introduced artificially through particular land uses and agricultural practices (e.g. nitrate is introduced extensively through fertilizer applications). Understanding the existence and role of all sources on their presence and distribution is essential to any assessment of water quality. In addition to being transported with the water, contaminants can undergo physical mixing and chemical and biological alteration within each pathway. Finally, it should be noted that chemical and biological activity may produce new contaminants that are potentially more toxic than the parent substance.

3.3 The modelling systems

3.3.1 THE PROBLEM

In order to address problems of the type illustrated in the introduction, hydrological predictions are required at appropriate spatial and temporal resolutions. The hydraulic and physical properties of the vegetation, soils, geology and stream networks can all vary substantially in space. Equally, the pattern of precipitation can vary significantly both in space and time. These fluctuations are important in defining the response of the hydrological system, and land-use change issues that are sensitive to such variations must be resolved at the appropriate scale. For example, a major concern is to identify the exceedance of a concentration threshold for nitrates, and since the major contributor to high concentrations in streams can be the flushing of nitrates in short duration overland flow events, then the modelling system must be capable of describing these flow events.

Unfortunately, in hydrological modelling such detailed predictions can be computationally expensive. Modelling systems such as SHE-TRAN require computational times of many hours on a fast workstation to run simulations spanning periods of several years, even for small river basins. This causes a conflict with the role of the DSS as an interactive utility. Users of the DSS normally require responses to questions within a matter of minutes, yet require information for land-use planning at the level of detail provided only by modelling systems such as SHETRAN. To overcome this conflict, the DSS implements NUARNO to provide 'broad-brush' hydrological predictions. The use of NUARNO allows the DSS user to explore the implications of a wide range of scenarios rapidly, and to choose specific hydrological scenarios for more detailed analysis with SHETRAN.

NUARNO is an extension of the ARNO modelling system developed at the Centro Idea, University of Bologna to model the surface water hydrology of the river Arno. It has been extended to permit predictions of the impact of future land-use patterns on hydrological responses at the river basin scale. The extended model provides rapid, low resolution predictions of groundwater and stream flows, as well as of stream water quality. The extensions to the model include a revised evapotranspiration component, the addition of subsurface groundwater and bank storage components and a stream water quality modelling component.

The SHETRAN hydrological modelling system is based on the SHE which was developed jointly by the Danish Hydraulic Institute, Sogreah (France) and the UK Institute of Hydrology (Abbott *et al.*, 1986a, b). It has been significantly revised and extended and can be used to simulate water flow, sediment transport and contaminant migration at a range of spatial scales from a single field plot to a large river basin (Ewen, 1990; Purnama and Bathurst, 1990).

The mathematical representations of flow and transport processes and the methods of solution used in the NUARNO and SHETRAN modelling systems are presented in the following sections.

3.3.2 NUARNO

(a) Flow

At the heart of NUARNO is a lumped, conceptual, soil moisture, accounting model, derived from earlier work carried out on the Xinanjian river system (Zhao, 1992) to represent soil water storage in a catchment. At a point, this model represents a soil compartment in which water can be stored. This compartment is connected to the main pathways for water movement as shown in Figure 3.2. Spatially, the set of soil compartments is described by the following empirical model of catchment soil storage capacity:

$$\frac{f}{F} = 1 - (1 - \frac{w}{w_\mathrm{m}})^b \qquad (3.1)$$

where
f/F is the fraction of the catchment in which the soil zone is saturated;

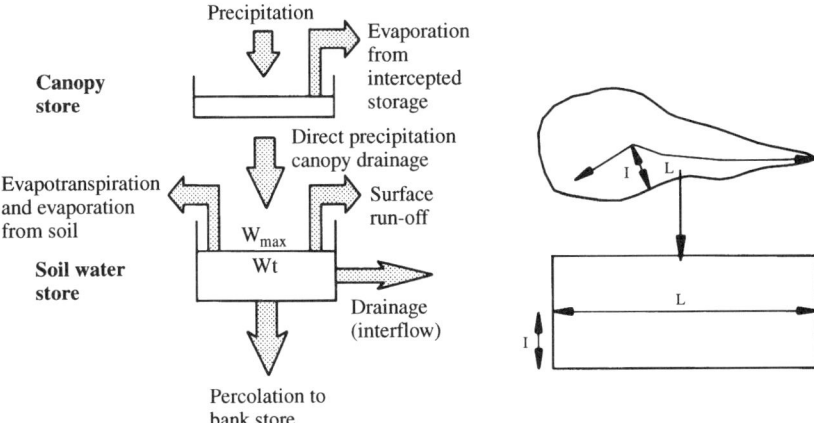

Figure 3.2 Schematic diagram of NUARNO model and transformation of subcatchment.

w is the cumulative soil moisture capacity in the soil zone;
w_m is the maximum moisture capacity in the soil zone;
b is a 'shape' factor.

The adoption of equation (3.1), known as a 'watershed storage capacity curve' (WSCC), permits the integrated storage, W, and the maximum storage, W_m, in the soil zone over a subcatchment to be defined. The coefficients w_m and b are calibration parameters that depend on the topography of the catchment and the soil characteristics. These coefficients have to be identified by fitting the model to the observed patterns of runoff using historical data.

Apart from the precipitation, P, all of the remaining flows in each of the main pathways are obtained as functions of the integrated soil water storage. Table 3.1 identifies each of the flow terms and the corresponding empirical model employed in NUARNO. The model for evapotranspiration has been taken from the SHETRAN system and uses an implementation of the Penman–Monteith and Rutter interception models (Dunn and Mackay, 1995). The processes of canopy interception, canopy evaporation, vegetation transpiration, direct precipitation to the bare soil surface, and evaporation from the bare soil surface are modelled.

The river basin is represented topologically as a set of subcatchments linked by a stream network. Each subcatchment is characterized as a rectangular region with a single stream channel along its central axis. To allow the prediction of changes in vegetation cover on the runoff from the basin, each subcatchment is divided into regions comprising hydrologically distinct vegetation types. In each region, the water released from the soil store to the groundwater drainage and overland

Table 3.1 Flow equations used in NUARNO

Process	Equation	Terms
Surface runoff R	$R = P - (W_m - W_t) + W_m \left[(1 - \dfrac{W_t}{W_m})^{\frac{1}{1+b}} - \dfrac{P}{(1+b)W_m}\right]^{1+b}$	P = Net rainfall W_t = Water content W_m = Maximum water content b = Curve shape parameter
Interflow D	$D = D_1 + D_{max} \left(\dfrac{W_t - W_{dtr}}{W_m - W_{dtr}}\right)^{d_{exp}}$	D_1 = Drainage rate at threshold value D_{max} = Maximum drainage rate W_{dtr} = Threshold water content for non-linear drainage d_{exp} = Exponent of drainage curve
Percolation to groundwater I	$I = I_1 + I_{max} \left(\dfrac{W_t - W_{ptr}}{W_m - W_{ptr}}\right)^{i_{exp}}$	I_1 = Percolation rate at threshold value I_{max} = Maximum percolation rate W_{ptr} = Threshold water content for non-linear drainage i_{exp} = Exponent of percolation curve
Lateral overland flow and groundwater routing	$\dfrac{\partial Q}{\partial t} + C\dfrac{\partial Q}{\partial x} - D\dfrac{\partial^2 Q}{\partial x^2} = Cq$	D = Kinematic wave diffusivity C = Kinematic wave celerity q = Lateral inflow x = Distance downstream Q = Flow, t = Time
Aquifer to channel Flow Q_B	$Q_B = \dfrac{\pm K_B L_C [H_B - H_C]^2}{2W_B}$	K_B = Bank permeability L_C = Length of channel H_C = Water head (channel) H_B = Water head (bank) W_B = Width of bank

flow components over any time interval, Δt, is assumed to be uniformly distributed in space. Using this conceptual simplification, the generated outflows from the soil zone are routed to the stream along a path that is orthogonal to the stream (Figure 3.2). Once the flows reach the stream, they are routed along the stream channel to the outlet of the catchment.

An analytical solution to the flow routing equation (Table 3.1) can be derived for a pulse injection. Numerical convolution methods are used to integrate the time series of outflow pulses, defined for each time step, from the soil zone in each vegetation region, to obtain the discharge at any point in space and time downstream (Adams et al., 1995).

In the groundwater zone, the effect of groundwater storage in the stream bank is accounted for by a stream bank component. This acts by buffering the groundwater discharge to the stream during high stream flows. Empirical equations are used to derive the height of the water in the stream from the predicted stream discharge and to calculate the change in groundwater storage in the adjacent stream bank.

(b) Transport

The NUARNO modelling system includes DYNUT, an in-stream water quality model, to permit the simulation of the fate of both point and distributed source inputs to contaminants in the stream system. DYNUT permits the simulation of the change in time and space of the following chemical indices: dissolved oxygen; biochemical oxygen demand (BOD); chloride; ammonia; nitrate; phosphate.

Components are being introduced to model the changes in pH and alkalinity due to both the carbonate systems in rivers and, if present, iron and manganese discharges to a river system.

The physical processes of advection and dispersion in the stream are modelled as well as the chemical growth and decay components for each of the chemical indices. The growth and decay components use relationships presented in USEPA (1985). The major relationships are presented in Table 3.2. Advection rates are defined from the flow component and dispersion coefficients (ε) are defined from experimental evidence or from the following empirical equation (Lin's formula):

$$\varepsilon = 0.32 n S_0^{0.25} \frac{Q^2}{U^{1.5} D^{2.75}} \qquad (3.2)$$

where
n is correction coefficient;
D is depth of flow [L];
U is velocity [LT^{-1}];
S_0 is longitudinal slope;
Q is discharge [$L^3 T^{-1}$]

Table 3.2 Transport equations used in NUARNO

Process	Equation	Terms
Contaminant transport	$\dfrac{\partial c}{\partial t} = \varepsilon \dfrac{\partial^2 c}{\partial x^2} - U \dfrac{\partial c}{\partial x} \pm S$	c = Contaminant concentration t = Time x = Distance along river U = River velocity ε = Longitudinal dispersion coefficient S = Sources and sinks of the contaminant
Surface reaeration	$P_R = \dfrac{21.7(9.14 \times 10^{-4} U)^{0.67}}{3.29 D^{1.85}} (DO_s - DO_t) \dfrac{\Delta t}{24}$	U = River velocity D = Depth T = Water temperature DO_s = Dissolved oxygen concentration at saturation DO_t = Dissolved oxygen concentration at time t Δt = Timestep
Photosynthesis	$P_P = P_{MAX} \sin\left(\dfrac{\pi(t - t_{sun})}{L_d}\right) \dfrac{BI}{D} \Delta t$	P_P = Increase in dissolved oxygen due to photosynthesis P_{MAX} = Maximum rate of photosynthetic oxygen production BI = Biomass of photosynthetic material per unit area t_{sun} = Time of sunrise L_d = Length of day

Nitrification	$R_N = AK6\, AM\Delta t$	AM = Concentration of ammonia $AK6$ = Denitrification rate coefficient
Biomass uptake	$R_U = BI\, AK7\, C\, \Delta t$	C = Concentration of ammonia, phosphorus or nitrate $AK7$ = Uptake rate coefficient
Plant respiration	$R_P = [\dfrac{AK5\, BI}{D}]\Delta t$	$AK5$ = Biomass respiration coefficient
Sediment oxygen demand R_B	$R_B = [\dfrac{BR}{D}]\Delta t$	BR = Sediment (benthal) respiration rate coefficient
BOD resuspension	$R_R = [AK1 + \dfrac{AK4\, U}{86400\, D}]BOD\,\Delta t$	$AK1$ = BOD first order decay rate coefficient $AK4$ = Resuspension coefficient BOD = Concentration of BOD

The stream network employed by the flow model is subdivided into reaches for the solution of the chemical transport equations. Finite difference approximations of the continuity equation for each reach are solved using the QUICKSORT numerical solution scheme. This scheme is stable and accurate over a wide range of in-stream conditions (Leonard, 1979).

Nitrate inputs from agricultural activities are derived in broadly the same manner as used in SHETRAN (section 3.3.3(b)). Other chemical information, including sewage effluent inputs, must be obtained from available data sources for the catchment of interest and are entered directly to the model as point inputs.

3.3.3 SHETRAN

(a) Flow

SHETRAN is a physically based, spatially distributed modelling system, which allows the spatial variation of the hydrological features of a catchment to be represented numerically. The representation of the spatial distribution of the catchment characteristics is provided through the discretization of the catchment into an orthogonal network comprising two types of elements – grid elements and stream channel links. The size of the grid elements is at the discretion of the user, but for full catchment simulations, a typical grid size of 1km × 1km is used, whereas for local simulations, covering a single farm for example, sizes of 100m × 100m may be adopted. At each grid element, the catchment is further subdivided into a vertical column of layers which represents the vertical paths for water and contaminant movement, and storage from the top of the vegetation canopy through the soil zone to the base of the underlying aquifer. The stream channel links are used to describe the movement of water and contaminants along the stream network in the catchment and are linked to both the unsaturated and saturated zones of the adjacent grid elements. Lateral flow may arise as either overland flow or subsurface saturated zone flow. The model does not include a representation of interflow in the soil zone, but does include stream bank components to represent storage in this zone.

Equations describing the balance of water within each element of the network are defined. The Boussinesq equation for two-dimensional groundwater flow is used for saturated subsurface flow (Bear and Verruijt, 1987) whereas Richard's equation describing unsaturated flow is used to model flow in the soil zone. Constitutive relationships are used to describe the water transfers between each of the elements and the storage in the model. The continuity equations for all elements are then transformed into systems of linear equations using finite difference methods that can be readily solved using standard matrix solution methods. The full set of element equations is established to conserve fluid mass and the equations are solved using explicit methods wherever possible to maximize computational efficiency.

The main equations employed in the representation of water movement in the canopy, the soil zone, the groundwater system and the stream network are presented in Table 3.3 and a visual description of the flow processes is presented in Figure 3.3.

(b) Transport

A nitrogen modelling system (NMS) has been developed to run within the framework of the DSS (Lunn *et al.*, 1996). This system is an extension of the general transport modelling approach employed in SHETRAN (Ewen, 1990). Three elements exist within the NMS. The first element forges the link between agricultural land-use model and the hydrological system. This element uses the Erosion/Productivity Impact Calculator (EPIC), an established crop growth and farm management model (Jones *et al.*, 1991). For a given land use and land management scenario EPIC provides data on the temporal distribution of nitrogen added to the soil system from agricultural activity as well as

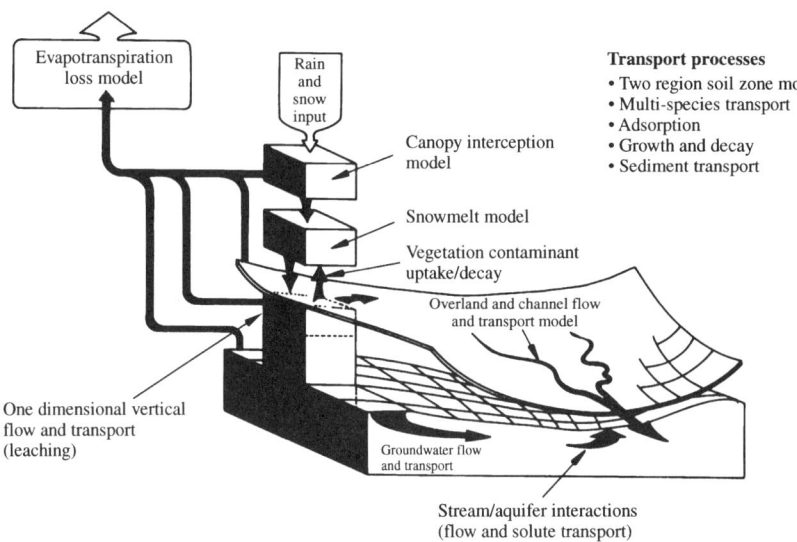

Figure 3.3 Schematic of SHETRAN flow and transport components.

Table 3.3 Flow equations used in SHETRAN

Process	Equation	Terms
Actual evapotranspiration	$$E_a = \frac{R_n \Delta + \dfrac{\rho c_p \delta e}{r_a}}{\lambda [\Delta + \gamma(1 + \dfrac{r_c}{r_a})]}$$	R_n = Net radiation (W/m^2) Δ = Rate of increase with temperature of the saturation vapour pressure of water at air temperature (mb/C$^\circ$) ρ = Density of air (kg/m^3) c_p = Specific heat of air at constant pressure (J/kg/C$^\circ$) δe = Vapour pressure deficit of the air (mb) r_a = Aerodynamic resistance to transport of water vapour from the canopy to a plane 2 m above it (s/m) λ = Latent heat of vaporization of water (J/kg) γ = Psychrometric constant (mb/C$^\circ$) r_c = Canopy resistance to water transport
Interception, storage	$\dfrac{\partial C}{\partial t} = Q - ke^{b(C-S)}$ where $Q = pp'P - pp'E_p\, C/S$ when $C < S$ $Q = pp'P - pp/E_p$ when $C \geqslant S$ $pp' = pp'$ when $p' < 1$ $pp' = pp'$ when $p' \geqslant 1$	C = Depth of water on the canopy S = Canopy storage capacity P = Rainfall rate p = Proportion of ground in plan view hidden by vegetation at its maximum extent p' = Ratio of total leaf area to area of ground covered by vegetation E_p = Potential evaporation rate k and b = Drainage parameters t = Time

Overland flow	$\dfrac{\partial h_o}{\partial t} = \dfrac{1}{A}\left[\displaystyle\sum_{i=1}^{4} Q_i + Q_R\right]$	h_o = Water depth A = Surface area Q_i = Lateral influxes Q_R = Vertical influxes
Unsaturated flow	$\dfrac{\partial \theta}{\partial t} = \dfrac{\partial}{\partial z}\left(D\,\dfrac{\partial \theta}{\partial z}\right) + \dfrac{\partial K}{\partial z} - E_s$	θ = Soil moisture content D = Diffusivity K = Hydraulic conductivity z = Depth E_s = Loss rate at depth z in the soil profile t = Time
Saturated flow	$S\,\dfrac{\partial h}{\partial t} = \dfrac{\partial}{\partial x}\left(K_x H\,\dfrac{\partial h}{\partial x}\right) + \dfrac{\partial}{\partial y}\left(K_y H\,\dfrac{\partial h}{\partial y}\right) + R$	S = Specific yield h = Water table elevation K_x, K_y = Saturated hydraulic conductivities H = Saturated thickness R = Recharge x, y = Space dimensions, t = Time

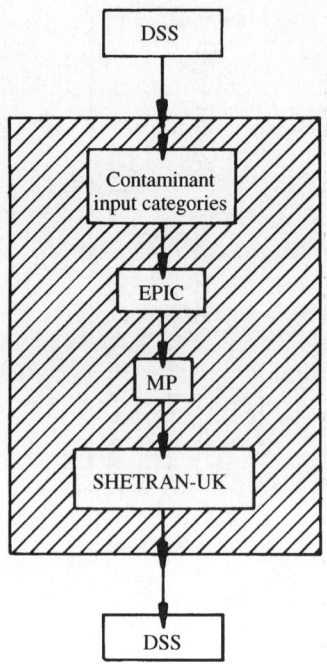

Figure 3.4 The nitrogen modelling system.

predictions of the time-varying uptake of nitrogen by the crops. Nitrate in rain is also added to the land but this is identified separately within SHETRAN. The second element comprises the modelling system MP (Lunn and Mackay, 1995) for predicting the distribution of nitrates in the soils beneath each different land cover. This component permits rapid prediction of initial conditions for input to SHETRAN. The third component comprises the SHETRAN transport modelling system, which predicts the migration and fate of nitrates throughout the river basin. The sequence of application of the model components within the NMS is shown in Figure 3.4.

The contaminant transport processes modelled in SHETRAN are illustrated in Figure 3.4. Contaminant transport within SHETRAN is calculated using a fully implicit, finite-difference scheme, defined for the same grid cells and stream elements adopted for the flow solution. Transport within the unsaturated zone is calculated using the convection–dispersion equation for one-dimensional vertical flow. In SHETRAN this model has been extended to include a dual porosity representation, in which water in the soil is partitioned between mobile and immobile zones. Lateral transport in the catchment can occur by advection and dispersion in the groundwater zone, in overland flow and in the rivers network. Chemical processes included in SHE-TRAN are linear and non-linear sorption, growth and decay, and multispecies transport of contaminants with non-cyclical interactions.

In order to keep computation time within reasonable bounds, nitrogen transport within SHETRAN is simulated using a single species model for nitrate which allows interaction with a pool of organic matter within the soil. Nitrates from fertilizer, animal waste and dead plant matter are applied at the surface and mixed within the root zone. Fertilizer nitrate added to the soil is dissolved in the soil water using the relationship:

$$\frac{\partial M_j}{\partial t} = -\frac{\partial(\theta c_j)}{\partial t} = -b_j\theta(C_j - c_j) \qquad (3.3)$$

where
j is contaminant species number
t is time [T]
M is mass of fertilizer per unit volume of bulk soil [ML^{-3}]
c is contaminant concentration [ML^{-3}]
b is constant rate coefficient [T^{-1}]
C is saturated contaminant concentration [ML^{-3}]
θ is moisture content

This allows gradual release of the fertilizer nitrate into solution as the soil becomes wet from a rainfall event. The uptake of nitrogen by plants and the decay of organic matter are both calculated using the EPIC simulation results. Denitrification (using a first order kinetic model of decay) can occur within the surface and subsurface migration pathways.

3.4 The modelling process

3.4.1 INTRODUCTION

Two types of land-use change scenario can be formulated within the DSS. First, land-use change may be generated by a new agricultural policy. In this case, the new distribution of land use is predicted by the Agricultural Economic linear programming (LP) model (Moxey *et al.*, 1995) in the form of percentages of each vegetation cover type per agricultural land class for each parish in the catchment. These percentages are then spatially distributed across the catchment using the distribution of the agricultural land classes and parishes. Second, a land use change may be defined by the user of the Decision Support System. This is achieved by selecting the required area on the DSS display, and applying percentage changes to the LandSat cover classes to define the required change. In either case, information is passed from the DSS to the hydrological models in the form of a new land cover distribution and an agricultural land-use map comprising vegetation cover percentages per kilometre square.

The first stage of the modelling process concerns the development of the base-line hydrological simulations. These must be representative of present day conditions. The base-line simulations must cover the whole catchment and must include a representative range of meteorological conditions, including both wet and dry years. This is necessary as specific effects of a land-use change may only be relevant to particular sequences of meteorological variation. In addition, the base-line simulations provide confirmation that the models are reproducing the hydrological behaviour of the catchment using historical data for validation (Figure 3.5). The second stage of the modelling process concerns the identification of the required outputs for the proposed scenario. The large volume of data that can be produced by the modelling systems prohibits the selection of all possible outputs. Consequently the user must carefully choose the spatial locations for output results and the types of data to be recovered from the simulation. This may be river discharges at selected key points but may also include, for example, soil moisture deficit distributions for selected months, or spatial distributions of groundwater level and groundwater nitrate concentrations. The third stage is to use the NUARNO modelling system to identify the scale of the change that might occur across the catchment. Finally, the fourth stage concerns the use of the SHETRAN modelling system to address the detailed hydrological changes for the most important scenarios. The third and fourth stages are discussed in more detail in sections 3.4.3 and 3.4.4. The output data from the models can be recovered by the user in a range of display formats including spatial maps, time series plots, flow and concentration duration curves and mass balance diagrams.

To illustrate the application of the hydrological modelling systems in the DSS, some basic results for the Cam river basin are presented. The Cam basin is a lowland, agricultural catchment located in East Anglia, UK, some 100 km north east of London. It has an altitude ranging between 168 m, in the chalk uplands to the south of the catchment, and 4 m in the fenland region to the north. Apart from the chalk uplands, which comprise low, gently rolling hills, the catchment is flat, with wide gently sloping river valleys draining northwards into the Great Ouse river. Figure 3.6 shows the three rivers, the main geological units and the seven subcatchments identified to model the basin. At the outlet of each subcatchment is a flow gauging station. Cereals, root crops and grasslands cover 90% of the area, with cereals the dominant crop in the south, and root crops, such as potatoes and sugar beet, dominant in the north. Rainfall over the catchment is low and fairly uniform across the region. There is a small variation in mean annual

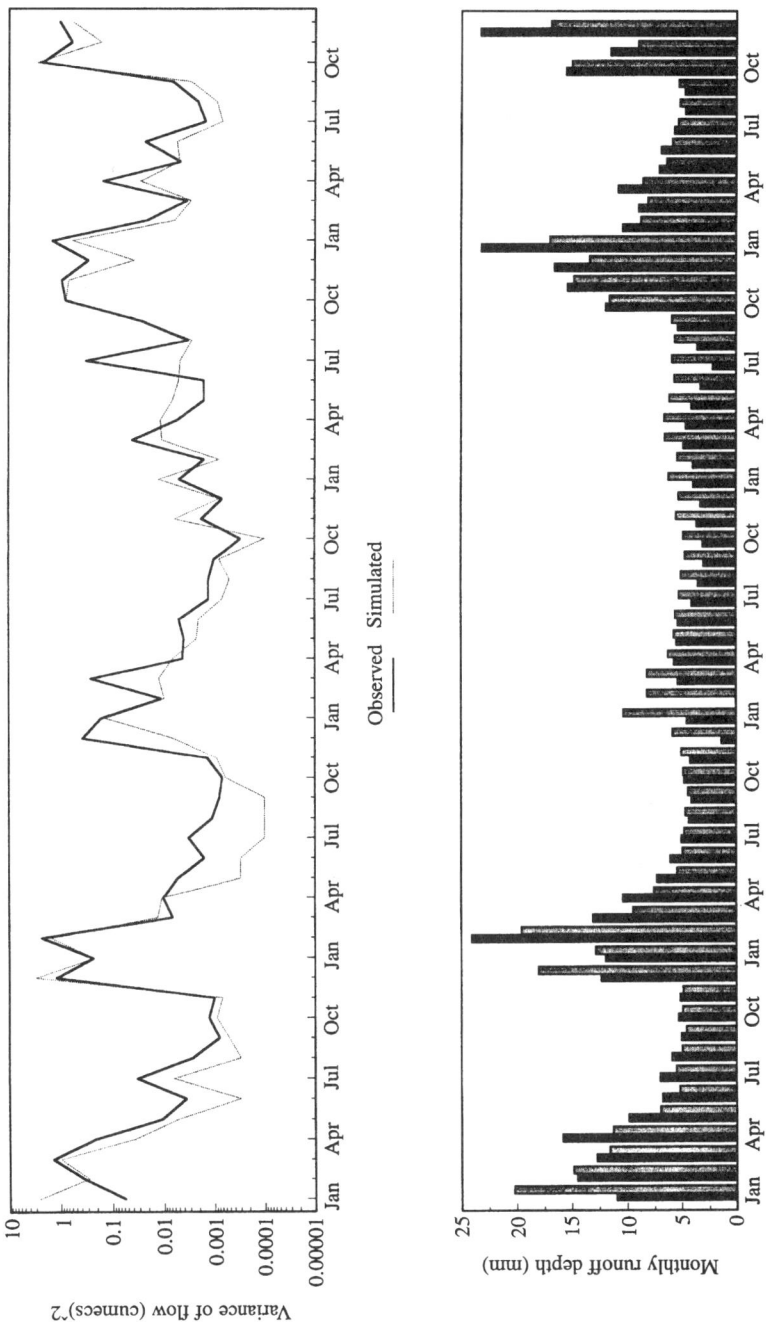

Figure 3.5 Cam catchment: monthly runoff and flow variance 1989–93.

60 Hydrological modelling

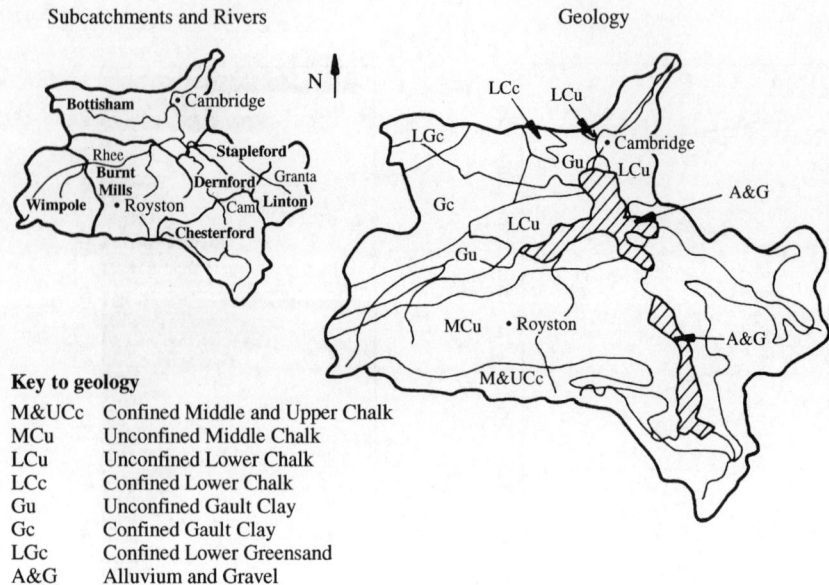

Figure 3.6 Geology of the Cam catchment.

totals from 500mm in the north to 650mm in the south. Meteorological data for the area also show a similarly uniform pattern. Water supply for public and industrial consumption is important in the region and the chalk aquifer has been heavily exploited. During dry periods, over-abstraction has caused low flows in several of the rivers of the catchment (Adams, 1994). This has had adverse effects both on the ecology of the rivers and on the use of water for seasonal purposes such as spray irrigation. Rivers tend to be shallow and slow-flowing and are maintained primarily by underground discharges. Little surface runoff occurs in the southern areas overlying the chalk.

3.4.2 THE DATABASE

As far as possible, national datasets are used to provide the spatial data for input to the two modelling systems. Topography is provided from the Ordnance Survey Digital Terrain Model but has been corrected by the Institute of Hydrology (IH) to be hydrologically consistent with their digital rivers network for the UK. The IH digital river network datasets are used to define the drainage network for the models. Vegetation distributions are provided by the LandSat remotely sensed database and soil association maps produced by the Soil Survey of England and Wales are used to define the soil distributions. The soil associations are converted into soil series, which identify the physical characteristics of the soils. Geological data are provided from the British Geological Survey maps of solid and drift geology and the available hydrogeological mapping for the area. Meterological data and precipitation data are obtained from the stations adjacent to and within the catchments. These data are available from the Meteorological Office and the National Rivers Authority. Agricultural activity data are based on the Ministry of Agriculture, Fisheries and Food (MAFF) agricultural census records. Abstraction and effluent discharge data are obtained from the consent licences held by the

Table 3.4 Data sets used for SHETRAN and NUARNO

Data set	SHETRAN	NUARNO
IH 50m digital elevation model	Catchment boundaries, grid element elevations, channel elevations	Catchment boundaries, channel slope
IH digital rivers network	Channel locations	Channel length
LandSat remotely sensed data	Vegetation distribution	Vegetation distribution
Vegetation parameter database	Vegetation parameters	Vegetation parameters
Soil survey soil association maps and soil series physical data	Soils distribution and physical parameters	Not used
Digital geological maps	Saturated hydraulic conductivities, depth of impermeable bed	Not used
Meteorological data	Hourly net radiation, vapour pressure deficits, temperature	Hourly net radiation, vapour pressure deficits, temperature
Precipitation data	Hourly precipitation	Hourly precipitation
MLURI land capability/ agricultural census data	Crop cover distribution	Crop cover distribution
EPIC plant nitrate uptake estimates	Nitrate uptakes	Not used
EPIC nitrate runoff	Not used	Nitrate inputs to stream
NRA discharge consent data	Not used	Point source discharge locations and loads
NRA abstraction licences	Ground and surface water abstractions	Ground and surface water abstractions

National Rivers Authority. The values of the parameters needed as input to the models are prepared by transformation of these spatial datasets using published point measurement data from the river basin and from general reference information. The transformation of the spatial datasets to the required form for input to the two modelling systems is carried out using a suite of transformation algorithms, collectively linked through a simulation management software system called SHE-SHELL. This system is used to provide the link between the data base, the DSS front-end and the two modelling systems. The advantage of this system is that it is entirely automatic and permits the user to carry out scenario analysis without concern for the development of the datasets needed as input to the models. Table 3.4 summarizes the primary datasets used by the hydrology models. The transformation algorithms were developed and tested on the Tyne Catchment (Adams *et al.*, 1995). All raw and processed data are stored in the DSS database and can be accessed directly through the user interface and by the models.

3.4.3 SCENARIO ANALYSIS: NUARNO

Five years of observation data are used to establish the baseline conditions. Five-year predictions are chosen as the shortest period consistent with providing a range of meteorological conditions and effects. These data must cover a consecutive sequence of years including both wet and dry periods. The NUARNO model is then used to predict the runoff at each of the gauging stations for this period. Calibration of the geometric parameters (w_m and b) of the NUARNO model and the drainage coefficients is required using at least part of the observation record. Preference is given to the use of only a part of the record since the comparison of the results with the remaining observation data provides a split sample validation of the model results. The results of the predictions are compared with the observations of actual runoff, and the components of the runoff hydrograph are analysed to show that the model and the real basin produce similar patterns of behaviour (Figure 3.7). The early part of the simulations is ignored since it is sensitive to inaccuracies in the initial soil and groundwater storage components of the model. Precise replication of the observations is not expected since errors in the input data as well as averaging in the model limit the accuracy of the models. However, the model must reproduce with reasonable accuracy the patterns of flows in each of the meteorological states (high, low and intermediate precipitation) and must reproduce the short time-scale (daily) variance in the discharge record. If both conditions are achieved then it is reasonable to

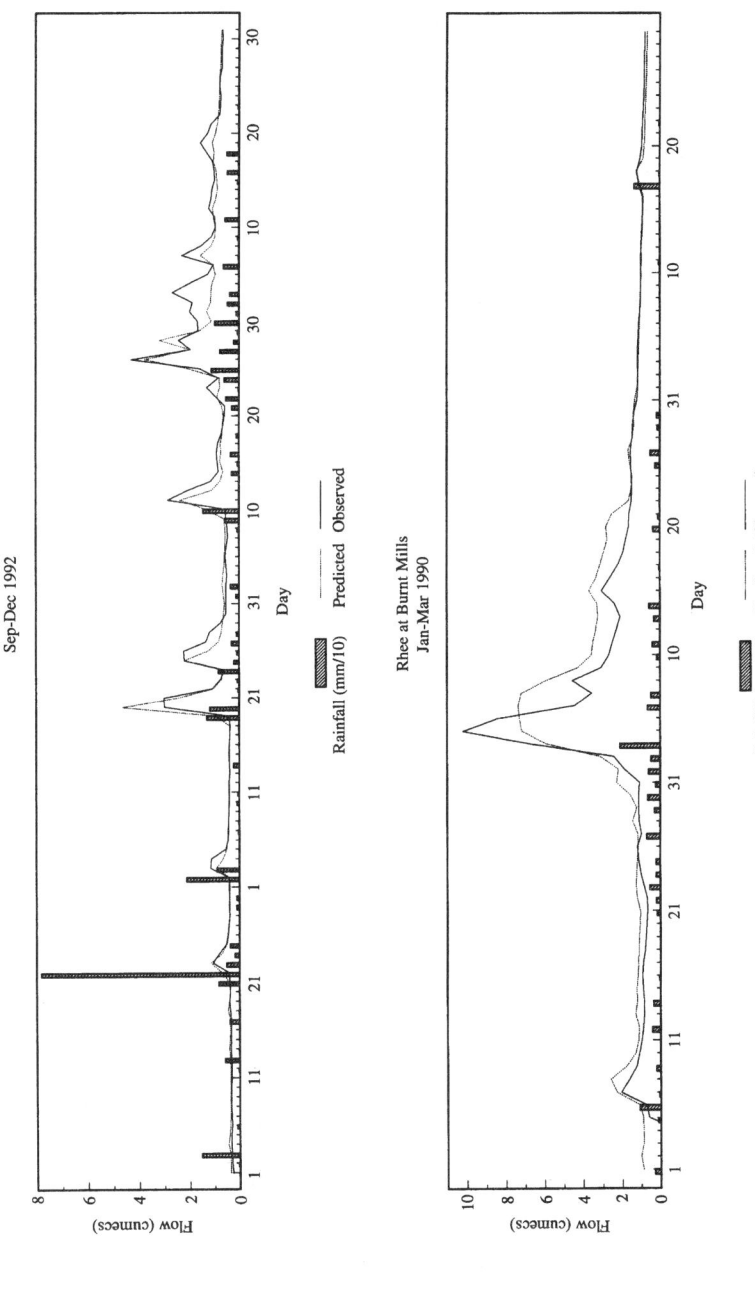

Figure 3.7 NUARNO predicted flows for Cam catchment.

expect the model will show the direction of both short and long time-scale changes in behaviour in response to a land-use change.

Once calibrated and validated, the model base-line results are defined and the model can be used for scenario evaluation. In all cases, scenarios are run for the same five-year meteorological record to provide a consistent basis for scenario intercomparisons. The outputs from the model are chosen in relation to the type of scenario and the issues considered to be important in the region. Issues may include summer base flow magnitudes in the streams, soil moisture deficits in the autumn prior to winter cereal sowing, or groundwater storage during low rainfall years. Other issues may require a knowledge of flood levels and the flushing of chemicals from waste effluent. In the Cam basin, each of these issues is important and therefore a range of data outputs are needed to interpret the magnitude and direction of any change in the water balance of the catchment.

Comparison of results is achieved using three primary tools, a water balance display (Figure 3.1), a stream-flow, discharge–duration plot and a contaminant concentration–duration plot. Time-series plots can also be displayed. Each tool can be chosen to show results for any given time period which allows the magnitudes of the effects for different seasons and years to be identified explicitly.

For those scenarios that show significant impacts on the river basin, detailed modelling must be performed using SHETRAN. Although the results from the NUARNO model are used primarily for screening scenarios for hydrological impacts, they may also be used to provide boundary condition data for input to local scale SHETRAN models. This extension of the use of NUARNO is discussed in the next section.

3.4.4 DETAILED ANALYSIS: SHETRAN

In the early developments of the DSS, the SHETRAN modelling system was used to model the entire river basin (Adams *et al.*, 1995). However, the computational complexity of this approach was found to be too great to be justified in all cases. Recent developments to the DSS now employ the SHETRAN modelling system in a way which exploits its modelling sophistication whilst avoiding excessive computational effort. In this approach it is assumed that the NUARNO model provides adequate descriptions of the regional pattern of change, but does not provide the fine scale or localized detail needed to support the economics and ecology models. The SHETRAN modelling system is used to examine the hydrological behaviour at the local scale for selected representative zones within the catchment (Figure 3.8). The identification of the zones depends on the problem being solved. The three examples presented in this book are concerned with the effect on

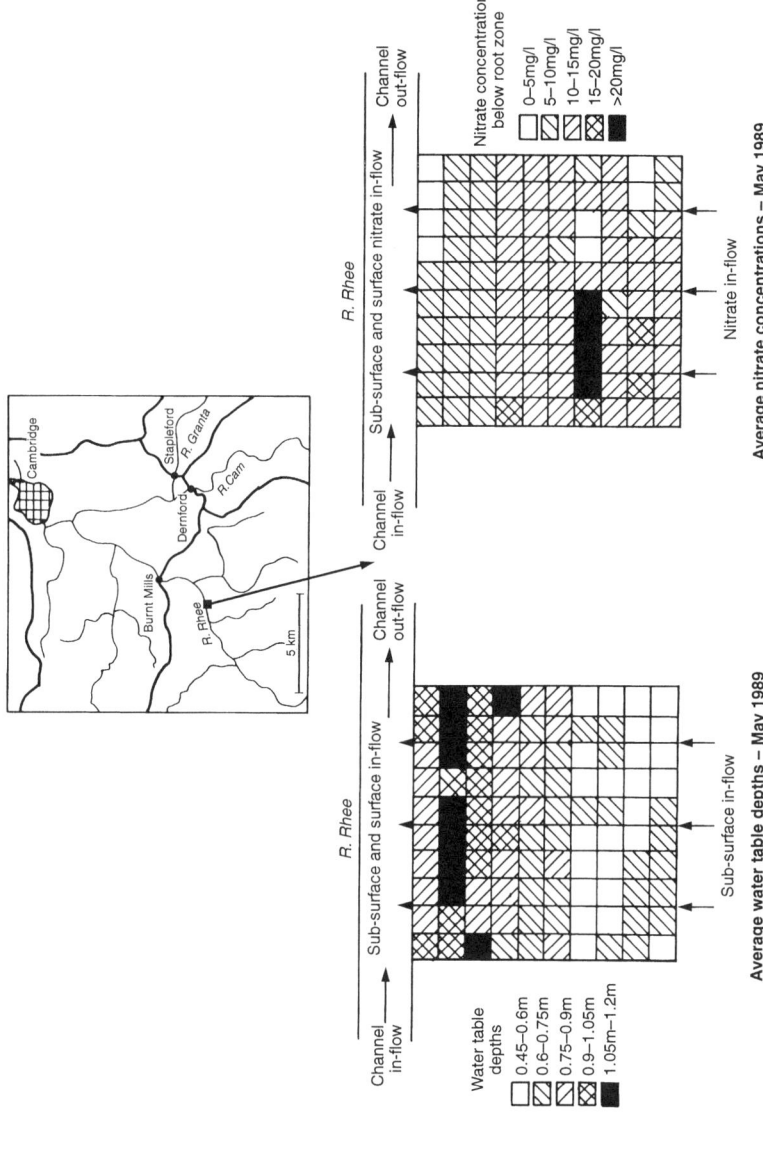

Figure 3.8 SHETRAN local model.

the local water balance of: (a) a change of management and use of the land along the river corridors in the Cam; (b) the use of relatively poor quality farm land for forestry; and (c) the effectiveness of imposing an irrigation tax. The GIS components of the DSS may be used to establish the areas of land to be altered as well as the general characteristics of the areas affected by the changes. Once the general characteristics of the change are quantified, representative zones that typify the proposed areas can be identified and a SHETRAN model developed for the chosen locations. The physical components of the SHETRAN models are obtained directly from the database and assistance in estimating the boundary and initial conditions may be obtained from the NUARNO system.

The interest in these simulations lies in how they quantify the hydrological impacts of a range of local land-use options on the available water resources. Taking the case of the river corridors as an example, the aim has been to quantify the changes in groundwater discharge to the streams in response to a change in land management on a narrow strip of land adjoining the stream. The size of the strip is to be determined and the land management practices on the land are to be compared for hydrological significance. The scale of these changes is far below the resolution of the basin scale NUARNO model and must therefore be tackled using SHETRAN.

Figure 3.8 illustrates the SHETRAN model as used for the river corridor study. Since flows to the stream are essentially at right angles to the line of the stream, the adoption of a rectangular domain for the SHETRAN model is acceptable. The model is run using the same basic procedure as for the NUARNO model. Base-line conditions are identified initially, and the results checked for consistency with the regional picture of water movements defined by the equivalent NUARNO scenario simulation. Once the base-line conditions have been constructed for the fine scale, sensitivity tests may be performed to identify the net effect of parameters, which are poorly known at the local scale. For the river corridors, the magnitude of the denitrification reaction rates is hard to estimate for the nitrates in the soil zone and the underlying groundwaters. Its impact on the results from the model must therefore be quantified. The sensitivity tests provide an indication of the reliability of the model results. Finally, the model is used to test the range of development, planning and management options that have been identified. The results can be displayed and compared for different scenarios using the standard output tools.

The results provided by the fine-scale modelling using SHETRAN are adequate for planning purposes. However, given the natural uncertainties in the database, additional data should be sought during any implementation phase to verify the conclusions derived from the simulations.

3.5 Issues in modelling

3.5.1 SCALE

Historically, SHETRAN and its predecessor, SHE, have been used to model small catchments of the order of several square kilometres in size, using a grid resolution of not more than 250 m. The modelling of the Tyne basin within the NELUP research programme (Adams *et al.*, 1995) was the first test application of SHETRAN to a much larger catchment using a coarse 1 km square grid. The application of the model at this scale raises a number of issues about the representation of the physical properties and processes governing flow in the catchment. Representations of both the topography and stream network reflect the size of the grid resolution so that at coarse scale considerable detail at the subgrid scale may be lost. In some cases, this loss of detail can lead to a loss of the information that is most relevant to land-use planning. For example, the effects of land drainage must be included in any model of the impact of upland forestry. This has been achieved in the DSS by the use of effective surface and subsurface parameters. The validity of this approach is strongly dependent on the characteristic behaviour of the small scale features. Dunn and Mackay (1996) show that effective parameters can be developed to reproduce large-scale flow behaviour of subgrid scale forestry drainage. However, the approach used is inappropriate for contaminant transport. The action of lumping processes together to yield a coarse scale model may only produce a model that is appropriate to a limited class of problems. The recognition of this restriction is essential if models are to be employed successfully at different scales.

In addition, it is recognized that within each square kilometre the spatial variation in soil and geological properties is great. To reproduce the aggregated response both at a point and over the river basin as a whole, it is necessary to identify appropriate kilometre-scale-effective transmissivities and soil parameters. The process of aggregation leads to predictions which must be interpreted with caution.

The problems of aggregation and loss of information also extend to the application of SHETRAN at the local scale. They are equally relevant to the development of a model of a catchment using the NUARNO modelling system in which averaging at the subcatchment scale is employed. In all cases, the issues to be considered in assessing the fitness for purpose of the models are twofold. First, the user must identify from the available data the values of the model parameters for any simulation. Second, the user must appreciate the extent of the loss of information and the bias in the results that can arise from the use of coarse scale averaging. The models within the DSS have been built with these considerations firmly in mind.

The application of NUARNO to provide a screening of different land-use changes relies on the assumption that the loss of information occurring in the use of this type of model is small enough to allow the identification of the direction of change in both contaminant transport and flow behaviour within the catchment as a whole. The application of the finer scale simulations using SHETRAN acknowledges the value of improving the level of detail in the model to yield the results required to quantify the extent of the impact of the land use change. The finer scale detail is needed to provide the quantitative data for planning.

3.5.2 PARAMETER UNCERTAINTY

In addition to the problem of scale, the issue of data and parameter uncertainty is important in the assessment of the utility of hydrological models. This is particularly important where the results from the modelling system are deterministic and no measure of the reliability of the results from the model is available. Spatial data for input to the models have been obtained as far as possible from available maps. However, the transformation of these spatial data into the spatial parameter distributions required by NUARNO and SHETRAN has been obtained from a limited set of punctual data (e.g. soil properties) and from representative data held in the published literature (e.g. vegetation canopy data). These data are subject to translation and sampling error as well as inaccuracy arising from the effect of spatial variability. These data must therefore be assumed at the outset to have limited reliability. The validity of the model datasets can be assessed through the validation process by a comparison of the output of the models with the available historical data. Although the results of the modelling are generally good for the Tyne (Adams *et al.*, 1995) and Cam basins, this is not a guarantee that the models are correct. Consequently, tests of the sensitivity of the response of the model outputs to errors are also needed to support their use. Full sensitivity analyses are inappropriate for models using systems such as SHETRAN as the number of degrees of freedom in the input data is vast. However, reduced analyses using procedures such as fractional factorial design can be adopted to provide the first order effects of uncertainties on the outputs of the system.

3.6 Conclusions

In this chapter the hydrological modelling components of the Land Use Programme DSS have been described and the approach to application of the modelling systems to examine land-use change impacts on water availability and quality have been presented. The requirement for rapid assessment of scenarios using the DSS has been addressed by the adoption of the NUARNO modelling system. This system uses an integral approach to the prediction of the pattern of runoff and stream water quality and provides fast simulation of catchment behaviour at a coarse scale. The results from NUARNO can be used to screen the different land-use scenarios efficiently and the worst case scenarios can be identified. Particular scenarios can then be analysed in greater detail using the SHETRAN modelling system to establish more accurately the specific behavioural changes of the hydrological regime to the proposed land-use change.

The hydrology models have been developed to produce a wide range of water-based information that is relevant to land- and water-use planning.

References

Abbott, M.B., Bathurst, J.C., Cunge, J.A. et al. (1986a) An introduction to the European Hydrological System – *Système Hydrologique Européen* (SHE), 1: History and philosophy of a physically based distributed modelling system. *Journal of Hydrology,* 87, 45–59.

Abbott, M.B., Bathurst, J.C., Cunge, J.A. et al. (1986b) An introduction to the European Hydrological System – *Système Hydrologique Européen* (SHE), 2. Structure of a physically based distributed modelling system. *Journal of Hydrology,* 87, 67–77.

Adams, R. (1994) A review of water management practices in the Cam catchment. *NELUP Technical Report No. 41.* October 1994.

Adams, R., Dunn, S.M., Lunn, R. et al. (1995) Assessing the performance of the NELUP hydrological models for river basin planning. *Journal of Environmental Planning and Management,* 38, 53–76.

Bear, J. and Verruijt, A. (1987) *Modelling Groundwater Flow and Pollution.* D. Reidel, Dordrecht, The Netherlands.

Dunn, S.M. and Mackay, R. (1995) Spatial variation in evapotranspiration and the influence of land use on catchment hydrology. *Journal of Hydrology,* 171, 49–73.

Dunn, S.M. and Mackay, R. (1996) Modelling the hydrological impacts of open ditched drainage, *Journal of Hydrology,* (in press).

Ewen, J. (1990) Basis for the sub-surface contaminant migration component of the catchment water flow, contaminant transport, and contaminant migration modelling system SHETRAN-UK. *NIREX Research Report NSS/R229,* NIREX, Harwell.

Jones, C.A., Dyke, P.T., Williams, J.R. et al. (1991) EPIC: an operational model for the evaluation of agricultural sustainability. *Agricultural Systems,* 37, 341–50.

Leonard, P.B.A. (1979) A stable and accurate convective modelling procedure based on quadratic upstream interpolation. *Computer Methods in Applied Mechanics and Engineering,* 19, 59–98.

Lunn, R. and Mackay, R. (1995) Solution of multispecies transport in the unsaturated zone using a moving point method. *Journal of Hydrology,* 168, 29–50.

Lunn, R., Adams, R., Dunn, S.M. and Mackay, R. (1996) Development and application of a nitrogen modelling system for large catchments. *Journal of Hydrology,* 174, 285–304.

Moxey, A.P., White, B. and O'Callaghan, J.R. (1995) The economic component of NELUP. *Journal of Environmental Planning and Management*, 38, 21–33.

Purnama, A. and Bathurst, J.C. (1990) A review of three features of sediment transport in stream channels: dynamics of cohesive sediment, infiltration of transported sediment into the bed and stream erosion. *NIREX Research Report NSS/R237*, NIREX, Harwell.

United States Environmental Protection Agency (USEPA) (1985) Rate constants and kinetics. *Formulations in Surface Water Quality Modelling*, 2nd edn. Report EPA/600/3-85/040, US EPA Environmental Research Laboratory, Athens, GA 30613.

Zhao, R.J. (1992) The Xinanjiang model applied in China. *Journal of Hydrology*, 135, 371–381.

Landscape ecology and land use

4.1 Introduction

The land cover of Europe has undergone radical changes over the course of the last 10 millennia. In the United Kingdom, palaeobotanical evidence indicates that land cover was dominated by forest following the retreat of the ice-mass of the last glaciation. Some 5000 years ago, human activity began to have a large-scale impact on the land cover with extensive clearances occurring during the bronze age. Much lowland forest was gradually turned over to pastoral agriculture. This clearance was only the beginning of what has become a progressive intensification of land use in Europe, which has accelerated greatly over the last 50 years. Whilst land began to be used more intensively, progressively more of it was also being brought into use. In some countries such as the United Kingdom pressures on land have been so great that the great majority of land is now used for some purpose and few areas of truly undisturbed habitats remain. The land resource is now considered to be a valuable commodity requiring protection and political recognition of this fact has resulted in many forms of landscape protection. In some cases this has meant a renewal of land management practices that had been abandoned and a deintensification of land use. In the United Kingdom this is illustrated by the recent widespread establishment of Environmentally Sensitive Areas following the Agriculture Act 1986. Here, areas which are perceived to be of historical, ecological and archaelogical interest, and under threat from potential changes in land use, are protected by the maintenance or reinstallation of land-use practices which were deemed to be responsible for their occurrence. In the case of the Pennine Dales Environmentally Sensitive Area for instance, haymeadow agriculture is protected by the imposition of strict fertilizer and herbage cutting regimes and landscape quality by the requirement of landowners to maintain characteristic features such as drystone walls. Although there may be considerable interest in conserving, or maintaining, the integrity of landscapes, quantifying the efficacy of landscape conservation schemes is complex. Gauging the success of simple procedures aimed at preserving individual features, such as the length of dry stone wall, is easy, provided that information concerning the correct disposition of the features in the landscape is available. Assessing the consequences of any management practice which involves the manipulation of ecological systems at these large scales is inherently more difficult because ecological processes in landscapes are highly dynamic and involve complicated multilevel spatial interactions. Spatial processes in even small-scale ecological systems have been comparatively understudied and are generally poorly understood. It is true to say that there is no established methodology for quantitative assessment of ecological processes at large scales and indeed much past research in landscape scale ecology has been anecdotal, qualitative and lacking in scientific rigour (Wiens, 1992).

4.2 Processes, scale and the ecology of landscapes

In order to be able to assess the effects of land-use practices on ecological systems in landscapes it is essential to understand the processes determining landscape structure. These can be grouped into three, namely: geophysical and geochemical, human–economic, and ecological (Perez-Trejo, 1993). The relative importance of each of these groups of processes in determining the ecological dynamics of landscapes is dependent on their spatial and temporal scales. Broadly speaking, there is a positive correlation between the time and spatial domains over which each of the different processes operate. The largest scales are those of the geophysical and geochemical processes such as tectonic activity and soil formation. These, together and in combination with meteorological processes, create the surface and substrates on which land use is practised and ecological processes occur. Human–economic processes, on the other hand, have space domains that may be very large, but the time domains over which they typically operate can be much shorter. This reflects the fact that the processes are dictated by human behaviour and population requirements. The geophysical and geochemical processes constrain those of the human–economic activities because they effectively dictate what humans can do with the landscape. The ecological processes within a landscape in turn are affected by all the other processes. These processes are effectively at the bottom of a hierarchy which is determined and constrained by processes higher up, which have slower dynamics. The relative importance of these 'constraining' processes in determining the ecology of landscapes will depend on how species interact with them. The form of the interaction will be determined, to a large extent, by the respective space and time domains of the individual species of organisms comprising the ecological landscape components. Interactions are strongest within any hierarchy at the same level and become weaker between other levels. At one extreme are populations of micro-organisms, which generally have small spatial and time domains of the order of millimetres and hours, at the other extreme are populations of species like large vertebrates and trees, with ranges of many kilometres and time domains in years. How these organisms interact with the overlying landscape structuring processes will be very different. Short-term small-scale variations in climatic features or land use may have devastating effects on the survival of a population of a species with small spatial and short time domains, but may have negligible effects on populations of a species which is more widely spread and slower growing. Species may escape the effects of some processes simply because they do not interact at appropriate temporal and spatial scales.

It is obvious that geophysical and geochemical processes are at such a high level in the hierarchy of landscape structuring processes that their dynamics rarely interact directly with those of ecological systems, they simply provide boundary constraints for the ecological processes. Human–economic processes, like land use, occur at temporal and spatial scales that either encompass or are similar to those of ecological processes. Temporally, agricultural and ecological processes are on the same scale since they are determined by the daily cycle of the sun and the periodicity of the seasons. The spatial scale over which other human–economic processes operate is more varied. At one extreme they may be dictated by policy instruments which can be implemented at the continental scale. A good example is the area of land used to produce arable crops in the European Union. This is influenced directly by a Union wide policy determining output from the arable sector of agriculture. At the other end of the spatial scale is the use to which individual parcels of land (i.e. fields) are put. The cropping regime imposed on particular fields may be influenced by the financial resources available to the farmer or by sociobehavioural factors such as 'tradition'. The range of dynamics of these constraining processes interact to create a spatial mosaic of different land forms. The individual land forms are either known as tessarae (Forman and Godron, 1986) or Landscape Response Units (Perez-Trejo, 1993).

From the viewpoint of the organisms within them, such landscapes will comprise areas of habitat of heterogeneous quality, embedded within a matrix of non-habitat. The form and spatial disposition of these habitats within the landscape will in part determine the presence of the organisms themselves. In some cases the suitable habitat will be distributed as blocks in the landscape which, depending on the relative abundance of habitat to non-habitat, may be isolated or connected. Where landscape features form greater than 60% of the total landscape area, they are likely to be well connected over large scales. In other cases suitable habitat will be distributed as long linear elements in landscapes, as in the case of rivers. These features are well connected, but usually represent a very small areal component of the landscapes in which they are found.

In order to quantify the effects of landscape structuring processes on the ecology of landscapes it is necessary to understand the effects that each of the different landscape-structuring processes have on the distribution and abundance of organisms and the habitats they occupy. Since geophysical and geochemical processes are slow relative to those of the organisms in the landscape, their dynamics may not need to be analysed explicitly. As a first step the geophysical and geochemical environment can be assumed to be static, simply providing boundary constraints for the other processes occurring in the landscape. The choice of methodology for investigating the relationship between land use and the ecological composition of the landscape is less easily defined because it will depend on the type of land use (and the land-use change) considered.

The use to which the land surface of Europe is put is generally restricted to a few human activities such as agriculture, silviculture, industry or urban development. The balance between these land uses is highly dynamic and rarely in equilibrium. Changes between these activities can have dramatic effects on landscape structure. When land is utilized for industry and urban development both the structure and ecology of the landscape are changed completely and in many cases irrevocably. In these circumstances the landscape is generally left depauperate ecologically, and any further change in land use to silviculture or agriculture is usually difficult and costly. Even if abandoned, land under urban or industrial usage will take many tens of years to regenerate to conditions where natural or seminatural habitats prevail. Changes in land use from agriculture to silviculture, in contrast, may often be reversible, but are no less dramatic in their effects on landscape structure.

Whereas the impacts of changes between these four major land-use activities are obvious and often dramatic, variation in how each activity is practised may have equally important effects on landscape structure and ecology. This is particularly the case for agriculture, because this form of land use is more dynamic than the others. The frequency with which the land surface is subjected to manipulation during the course of agricultural activities means that such landscapes are highly disturbed and are generally unstable ecologically. Consideration of the changes arising during the course of a simple pasture to arable cropping rotation illustrates this. Change from pasture to arable will involve destruction of the grassland sward; disturbance of the soil ecosystem; probable use of fertilizers and pesticides. These procedures will result in the complete loss of plant communities that may contain as many as 30 species per metre square. They will also destroy associated invertebrate communities. In the case of spiders, grassland may contain as many as 150 species, most of which will be lost and replaced by a poor community with typically less than 20 species on the change to an arable habitat. At a larger scale the changes will also influence the composition of the soil fauna, leading to the disappearance of species that do not tolerate soil disturbance, as in several species of earthworm. The changes will also result in the loss of nesting sites for some species of birds. In addition, the use of pesticides and fertilizers will influence the species composition of the newly established arable habitat. These changes arise from just one type of land-use pattern, but the range of disturbance in other land management practices in modern agriculture is equally great.

It is clear that assessing the ecological impacts of changes in agricultural land use will be very complex. It is also obvious that there is very little scope for undertaking replicated manipulative experimental approaches at the landscape scale. Devising experimental procedures capable of analysing ecological processes and their responses to land use change cannot be undertaken for more than a restricted range of land use scenarios. It is also unlikely that the land resources or the means of undertaking experimentation could be made available for such research. A more practical approach is to develop modelling methodologies which can effectively simulate the landscape and landscape processes as databases and numerical algorithms. Experimentation can then be undertaken through manipulation of the databases and simulation of the different processes on a computer. It should be pointed out that modelling procedures of this type require not only a detailed knowledge of landscape structure and functioning, but also large amounts of data.

4.3 Investigating the ecology of landscapes

Organisms, whether as individual species or collectively as assemblages or communities, are the most important ecological components of landscapes. In order to assess the effects of land use and landscape structuring processes on the ecology of landscapes it is therefore necessary to understand the processes that determine the distribution and abundance of species. There has been considerable past research investigating these processes for a wide range of organisms, but there appears to have been almost as many approaches as there have been studies. Although the range of approaches used has been large, they can be separated into two broad streams. In one group are associative models that attempt to relate the distribution of individual species, or groups of species, to habitat and other environmental features within the landscape, and, in the other group, approaches that derive distribution patterns from models of the underlying life history processes that determine them.

4.3.1 ASSOCIATIVE APPROACHES FOR PREDICTING THE DISTRIBUTION OF SPECIES AND COMMUNITIES IN LANDSCAPES

All associative approaches rely on linking the distribution of the feature of interest (species or communities) to measures of explanatory variables, typically habitat, land use or environmental features. Having once identified suitable explanatory variables, their distribution can then be used to predict the distribution of organisms in other landscapes. Where the distribution can be linked to features that are altered by land-use change, it is a simple matter to predict the consequences of land-use change on that distribution.

(a) Aspatial associative approaches

There are very many techniques available for associating distributions with habitat/environmental features. At the simplest these linkages can be made using a 'rule base' approach, with no assumptions made about the statistical nature of the species/community distributions or the underlying explanatory variables. Thus, the observed distribution of the species/community is overlayed and compared with maps of different sets of environmental data. Coincidences between the observed distribution and the measured environmental variables can then be used to create a rule base defining where the species/community occurs in relation to the known distribution of the key environmental variables. The landscape is thus classified into areas with, and without, the species/community of interest. The approach can be extended from the categorical present–absent rule-base to include ranges of habitat suitability. This can be based on measures of the proportion of specific habitats occupied by the species/community within landscapes. Thus if 80% of all occurrences of a species are recorded in one habitat and 15% and 5% of records occurred in two other habitats, then the suitability ranking of the three habitats would be in the ratio 80:15:5. This approach has been used to identify land suitable for species and is particularly favoured by workers with access to Geographical Information Systems in which map overlay facilities are standard functions. Aspinall (1993) used it to predict the distribution of habitat suitable for red deer. Simple overlay analyses do not provide measures of the likely errors involved with their predictions. More sophisticated versions of this approach rely on Bayesian inference, deriving an estimate of the probability of encountering species/communities from assessments of their habitat requirements and measurements of the occurrence of these in the landscape. Griffiths *et al.* (1993) describe this methodology as a five stage process.

1. The habitat suitability of a site is defined by the presence or absence of the species. (The habitat characteristics found on sites where species are present are identified and considered as potential discriminators);
2. The significance of each habitat characteristic for discriminating between sites with and without the species is assessed using χ square;
3. The frequency of suitable habitat characteristics in the landscape is used to calculate conditional probabilities for each habitat characteristic;
4. The conditional probabilities are combined into a single posterior probability of habitat suitability for each site using Bayes' theorem;
5. The habitat suitability for sites with known habitat characteristics is predicted.

These authors used this methodology to predict the spatial distribution of upland birds in the northern Pennine region of England and southern Scotland. The estimation of posterior probabilities using Bayes' theorem assumes that the conditional probabilities are independent. This means that if the habitat characteristics or predictor variables believed to be responsible for determining the presence of a species are related in some way, then the estimates of habitat suitability derived from the application of Bayes' rule will be biased. Since it is likely that many habitat features in a landscape are likely to be correlated spatially this bias is likely to be present to some extent in most applications of any technique which utilizes spatial data.

Non-independence in predictor variables can be accommodated, in part, with more sophisticated associative modelling approaches, such as discriminant analysis. Discriminant analysis describes any activity connected with the task of classifying unknown objects into groups. The term has often been used rather loosely, but most techniques explicitly incorporate covariation in potential predictor variables in the procedure. The basic idea behind the approach in the context of landscape ecology is to identify suites of explanatory variables that can be used to identify sites/areas that are likely to have or not have species or communities. The simplest technique involves six stages:

1. Identification of groups into which objects are to be classified;
2. Identification of predictor variables for discriminating between the groups;
3. Estimation of the means of each predictor variable in each of the groups;
4. Estimation of the covariance matrix describing within and between variable variation in predictors;
5. Estimation of the separation of the object to be classified from each of the group means in terms of the measured predictor variables.

The distance measure usually used is the Mahalanobis distance which is an euclidean distance scaled by the covariance matrix;
6. Allocation of the object to the group from which the separation is smallest.

Chatfield and Collins (1980) and James (1985) provide the statistical background to Discriminant Analysis and Williams (1983) reviews its use in ecology. One of the major problems with using this approach is that it assumes that the covariance matrices for predictor variables in each of the groups are equal. In practical terms this means that the variation both within and between predictor variables is assumed to be the same over all groups. Violation of this assumption renders the discriminant procedure inaccurate. Although there are techniques available for assessing the validity of this assumption, these have only occasionally been used by ecologists, and generally the problem has been ignored.

Discriminant analysis has been used most frequently in ecology as a descriptive tool, with the object of identifying optimal separation of groups. Rushton *et al.* (1989) used it in this way to investigate the effects of pasture improvement found on the ground beetle faunas of upland grassland and Hill *et al.* (1993) to investigate the factors determining the wading bird communities and management on British estuaries. The most useful applications of discriminant analysis have been predictive, where the object was to provide methods for predicting the group to which an observation belongs. In this context the analysis can be used to identify the suite of environmental or habitat variables that distinguishes between points in the landscape which do and do not have an individual species or community. Moss *et al.* (1987) used this approach to predict river invertebrate communities from measures of environmental variables, and Rushton and Eyre (1992) to predict spider communities on the basis of soil wetness, site altitude and vegetation height. In both studies changes in the input variables could be used to predict the effects of land-use change on individual species distributions because many of these were directly linked to land use. Despite offering considerable potential for landscape research the use of discriminant analysis in landscape ecology has not been widespread.

Regression modelling techniques provide the most sophisticated of the associative approaches both for predicting the distribution of species or communities and the effects of land use change. In common with other associative approaches they utilize sample data on the distribution of the species/community of interest and relevant environmental or habitat characteristics measured at the same points in space. These techniques attempt to isolate systematic from random variation in the observed distribution of the ecological feature modelled, with the systematic component, a set of explanatory variables, explaining the response. The simplest statistical models have a systematic component linking the observed distribution of the species to features of the environment, and a random component which is a probability distribution used to explain variation in the response variable not attributable to the explanatory variables. It is possible to model incidence or abundance of the species or community of interest, depending on the type of data available for analysis. These models are one form of a family of related approaches collectively termed generalized linear models (McCullagh and Nelder, 1983). Choice of the 'correct' model for the non-systematic or random component of such models is critical to their validity. Incorrect specification of this component may render the model and subsequent predictions invalid. Where the distributional data are simply presence or absence, at each point in the landscape, logistic regression provides an appropriate model. Logistic regression has been used to investigate the relationship between species incidence and habitat features for many taxa. Rushton (1991) and Luff *et al.* (1991) used it to investigate the habitat preferences of linyphiid spiders and ground beetles, respectively, and Rushton *et al.* (1990) to investigate the effects of land management on the incidence of ground beetles. In these cases the analyses were undertaken assuming a binomial error structure within the generalized linear-modelling package GLIM. Similar approaches have been used to predict the distribution of trees (Lenihan, 1993) and birds (Buckland and Elston, 1993). Where data on counts are available, as in many population studies, models based on Poisson error structures may be more appropriate. Rushton *et al.* (1994a) used this approach to analyse the effects of habitat features on the abundance of bird territories along river corridors. Methodologies for assessing the adequacy of different error distributions in linear models are considered in Aitken *et al.* (1989).

Whilst there may be a clear statistical rationale for selecting the random or error component of linear models, choice of the systematic part is more complex. There are two main issues: first, the choice of explanatory variables and second, the form of the response curve. Ecological processes are not modelled explicitly in associative models, so it is assumed that the variables used as independent predictors are truly causal. A certain degree of foreknowledge is therefore required on the part of the ecologist to ensure that only causal variables are incorporated in the model. At their simplest, such models utilize habitat features known to be determinants of species or community distributions. The presence, or absence, of key habitat features, such as suitable sites for building nests for birds, are obvious choices for such models, because they have a direct effect on the incidence of the species themselves. These variables can be included as categorical or dummy variables. In many species, incidence and abundance may be dependent on the presence of more than one habitat feature and there may be interaction between explanatory variables. Identification of suitable variables for inclusion in such models then becomes more complex. In some cases it may be possible to combine explanatory variables, or reduce their number, through data reduction/ordination techniques like Principal Components Analysis (PCA). PCA and other ordination techniques reduce the dimensionality of complex datasets abstracting the major axes of variation within them, providing a set of new variables which are linear combinations of the original variables. These axes can then be used as explanatory variables in regression models, with species incidence or abundance as response variables. Rushton *et al.* (1991) used Reciprocal Averaging (RA) ordination to simplify grassland management characteristics into two axes of variation for modelling the responses of ground beetles to land-use change. Rushton *et al.* (1994a) used Detrended Correspondence Analysis to simplify 74 categorical variables describing river habitat structure into two axes for use as an explanatory variable explaining riparian bird distribution. Although ordination procedures may appear to offer a quick and easy way of simplifying suites of potential explanatory variables in the development of models for predicting the distribution of species, these approaches should not be used uncritically. Where there are obvious major ecological trends in the dataset analysed, these will be recapitulated in one of the first axes of the ordination. The simplification may then provide a useful summary of variation in key environmental attributes. In effect many of the key variables will be correlated. Where there are no trends in the data then the new variables may well have little ecological meaning. It is also worth noting that the values of the new variables derived from ordinations are dependent on how the variables are distributed within the dataset. Where there are many variables which are recorded rarely, the derived new variables may be prone to domination by these. Thus, an undue weight may be given to what is, in effect, a non-significant feature in the derivation of the new explanatory variable.

Selecting the correct response curve relating the occurrence of the dependent variable to the explanatory variables can also influence the utility of regression models. The simplest models utilize a monotonic response of the dependent variable to the explanatory variables. In these it is assumed that the incidence/abundance of the features of interest rise or fall in relation to the explanatory variables. These simplistic models may not necessarily be the most realistic. Jongman *et al.* (1987) detail methodology for fitting symmetric unimodal curves to incidence and abundance data. They argue that unimodal responses are more ecologically meaningful than simple monotonic responses, because many species show optima along environmental gradients. These techniques have been used to investigate species habitat relationships in a variety of taxa including plants (Austin *et al.* 1984), trees

(Lenihan, 1993), and invertebrates (Rushton, 1991). Although unimodal responses have a functional appeal as response models, such relationships need not necessarily be symmetric and as such they may not necessarily be any more realistic than monotonic responses. Austin *et al.* (1990) have demonstrated that species responses to environmental gradients whilst being unimodal may be skewed. Austin and Gaywood (1994) suggest that the direction of skewness may depend on where the species is on the environmental gradient modelled. They fitted truncated polynomial models to produce skewed models for species showing this type of response. Complex non-symmetric response curves can also be fitted using Generalized Additive Models (GAMs) (Hastie and Tibshirani, 1990). These are non-parametric extensions of Generalized Linear Models that explicitly utilize smoothing components derived from the response variable itself within the model. Yee and Mitchell (1991) have reviewed GAMs and their potential for use in evaluating species response curves to environmental gradients and have used them to predict the large-scale distribution of trees in New Zealand.

(b) Associative approaches that include spatial information explicitly

Most of the discussion above has centred on modelling methodology which has not considered space implicitly. Simple regression models may be expanded to include spatial aspects of habitat features as explanatory variables. Typically, these models include measures of variables local to the point at which the observations of the dependent variable were made. These approaches have been used extensively to investigate the role of habitat and landscape structures such as corridors and habitat fragmentation in determining the distribution of animals. Much of this work has been involved with mammals for which habitat measurements can be easily defined. Woodland size and distance to nearest woodland have been shown to be important factors determining the incidence of red and grey squirrels in landscapes (Fitzgibbon, 1993; van Appeldoorn *et al.* 1994). Similar studies have shown that the abundance of dormice is related to the size of ancient woodland, distance to adjacent woodlands and the number of hedgerows radiating from them (Bright *et al.*, 1994).

Regression models may be made more sophisticated by the use of spatial features which link the distribution and/or abundance of the modelled feature to the distribution of the explanatory variables in space. Spatially lagged models (Upton and Fingleton, 1985; Haining, 1990) effectively incorporate, as explanatory variables, measures of variables from the areas adjoining as well as those at the point of interest. These are termed models with spatially lagged explanatory variables. In these models the adjacent sites interacting with each point in space are enumerated in a weighting matrix (W) and a coefficient (ρ)

that quantifies the strength of the spatial interaction which is estimated in the regression. The form of the weighting matrix can have considerable implications for the derived regression since it determines which points in space interact. Choice of weighting matrices is considered in Upton and Fingleton (1985). The inclusion of space also has implications for the response variable in the regression model. The value that a response variable has at a point in space may also be related to the values of those around it; often being similar to values at points nearby but which become increasingly dissimilar with increased distance away. The response variable is then correlated in space (spatially autocorrelated). This means that measures of how the response variable varies spatially have to be incorporated into the model. As with the spatially lagged explanatory variable models a weighting matrix is used to define which sites interact spatially and a second coefficient is used to quantify the size and direction of the autocorrelation. Upton and Fingleton (1985) describe different methods for incorporating error structures of this type into regression models and Haining (1990) describes different model structures for the spatial variation component.

Both lagged explanatory and lagged response models effectively subsume elements of variation that may be considered to arise as a result of spatial ecological processes being brought into the model. In the former case, including measures of habitat features as spatial lagged explanatory variables incorporates the effects of habitat areas on species/community incidence or abundance. Spatially lagged response variables effectively incorporate the effects of the surrounding population into the model. Although these techniques would appear to offer considerable scope for investigating the effects of land-use change on the ecology of landscapes they have not been used extensively.

In contrast with the other associative methods described, statistical models have the advantage that they provide estimates of the accuracy of their predictions. All associative modelling approaches, however, share one major drawback in that they are essentially static and cannot be used to predict species distributions outside of the range of habitat conditions from which they were derived. This is a considerable restriction for studies of landscape dynamics because it presupposes that the consequences of future landscape structure scenarios can be predicted on the basis of those used to create the model. This assumption may be unrealistic. The development and implementation of set-aside as a means of controlling arable production within the European Union is a good example of this problem. Set-aside is a habitat that, in the 1990s, forms a major component of the lowland landscape in Europe and yet 10 years previously, the habitat did not exist. Associative approaches effectively subsume life history processes into correlations between species incidence and habitat variables and preclude organisms from adapting or changing within the landscape. Wiens (1989) concluded that predators and competitors, spatial and temporal changes in the type of habitat utilized, time lags in organism responses and degree of habitat saturation all limit the use of associative approaches in predicting bird distributions. These factors will influence bird distributions because they have effects over and above those arising from the presence of suites of necessary habitat characteristics. In other words, whereas key habitat variables may be necessary for the presence of a species, actual presence may depend on other ecological processes. It is likely that these ecological processes will play some part in determining the distribution of most species of organism.

4.3.2 DETERMINISTIC MODELLING APPROACHES FOR PREDICTING THE DISTRIBUTION OF SPECIES AND COMMUNITIES IN LANDSCAPES

Although associative models have undoubted simplicity, they lack ecological functionality. For the ecologist, a more functional approach for predicting the distribution of organisms would be to consider the processes occurring at the level of the population of the individual species. If population processes can be modelled within the landscape, then the distribution of the organism can be derived from the net outcome of the processes at each point in space. Models of population dynamics have an extensive history in ecological research.

The most flexible and robust population models have been those which have been developed specifically to test general ecological principles. The models of Lotka (1925) and Volterra (1926) that were developed to analyse interspecific interactions are the earliest of a series of theoretical approaches developed over the last 50 years. These approaches have been based on numerical analyses of differential equations. In the first population dynamics models individual life history processes like fecundity and mortality were not considered explicitly. Although these modelling approaches have undoubtedly been of use in generating ecological theory, their use in landscape ecology is limited. One of the main restrictions on their use is that by being so general there is often no obvious link between the model state variables of population growth rates and the landscape in which these processes are occurring. The development of metapopulation dynamics theory (Levins 1969) and subsequent metapopulation dynamics models have extended population modelling into the spatial domain. Metapopulation dynamics models analyse changes in the distribution of populations of species in terms of probabilities of extinction and colonization of habitat patches. By recognizing that populations exist in spatially heterogeneous environments these models go some way towards being of practical use in landscape scale research. Whereas the mathematical basis of the metapopulation dynamics model is elegant and the concept has intuitive appeal, unfortunately the existence of species populations in a balance between extinction and colonization is not a widespread occurrence (Harrison, 1994).

Metapopulation dynamics theory has been used to investigate the spatial dynamics of real species. This approach was used by Verboom *et al.* (1991) to investigate the dynamics of badger populations in fragmented landscapes in The Netherlands and by Okubo *et al.* (1989) to investigate the spread of grey squirrel in Britain. In both cases space was an important feature of the model but in neither study were population dynamics related to real landscapes. Application of metapopulation dynamics modelling to real landscape problems has not been widespread. Thomas and Jones (1993) related extinction and colonization probabilities in populations of the butterfly *Hesperia comma* to the spatial distribution of patches of suitable habitat. They then used the derived equations to predict the future distribution of the species in south-east England. Hanski and Thomas (1994) extended this approach to two further butterfly species *Melitate cinxia* and *Plebejus argus*, the former in Finland and the latter in the UK. A similar probabilistic, but non-metapopulation dynamics approach to predicting population spread was used by Usher *et al.* (1992) to describe the spread of the grey squirrel and decline of the red squirrel in England. In this study Markov modelling was used to predict the relative proportions of 10km squares in mainland Britain occupied by both species on the basis of past colonization and extinction rates.

More realistic population dynamics models link individual life-history processes to environmental features in the landscapes in which the organism is found. These models tend to be much more complex because they adopt a much more reductionist approach. Models of this type have been developed extensively in the past, but very few have been spatially articulate. These models require a detailed knowledge of the effects of all biotic and abiotic factors on the dynamics of fecundity, mortality and migration within the individual population. Even for a simple population model investigating the effects of only a few variables, the processes that need to be modelled can be large. Rushton and Hassall (1987), for instance, investigated the effects of just two variables, temperature and food availability, on the dynamics of a population of an invertebrate species and they included equations for a total of 11 processes, ignoring many more because of lack of data. The more detailed deterministic population models become, the more they lose their generality. In the case of the invertebrate population dynamics example described above, application of the model was restricted to grasslands where food resources were found. Whilst it may be comparatively easy to study the responses of population processes of survivorship and growth to environmental factors such as food availability and climate, linking these processes to the landscape necessitates consideration of dispersal and migration phenomena. Quantifying migration and dispersal in animal and plant populations is difficult, because both phenomena are usually highly stochastic. As a result of this, few population models have considered migration.

Although spatially explicit population dynamics models have not been much used in studies of animal populations they have been used extensively in plant ecology, particularly in species for which intraspecific competition for space may be an important determinant of population dynamics. Kenkel (1991) reviewed the major spatial approaches to modelling intraspecific interactions in plants. These models analyse life-history processes of individual plants within a population in relation to their proximity to each other. Plants are assumed to compete for light, nutrients and water with their neighbours. They are often termed neighbourhood models (Pacala and Silander, 1985). In a manner similar to the lagged explanatory variable models described above, plants are assumed to have greater effects on the life history processes of those plants which are nearest. The models quantify the spatial effect in terms of either zones of influence, which may be straightforward functions of distance, or tessellations of the space in which the population occurs. Tessellations are effectively a classification of the spaces that allocate areas of resource to each plant within the population. In both methods the extent of influence of plants on their neighbours may be weighted by the size of the plants. As with spatially lagged regression models, there is considerable debate as to the magnitude and range of the zone of influence. The use of tessellations makes the definition of potential interactions between individuals easier to define and Kenkel (1991) discusses evidence to suggest that much of the interaction between plants occurs between those sharing tile boundaries. Much of the research utilizing these approaches has been concerned with investigating intraspecific competition and has been undertaken with comparatively small fast-growing species at plot scales (e.g. *Arabidopsis thaliana* (Pacala and Silander, 1985)). The use of these approaches in landscape-scale research has been restricted to forestry, where measurements of individual performance have been used to provide information on likely forest yield.

The development of similar, spatially articulated, population dynamics modelling approaches for animals in real landscapes has not been extensive. Where this has been undertaken it has usually been fostered by conservation or population control requirements. In many cases modelling approaches have been followed in order to try and explain the decline of conservable species or the spread of pests. Lahaye *et al.* (1994) developed a spatially articulated population dynamics model for the endangered spotted owl in California. The model was used to investigate the future trends in populations of this species. This bird is restricted to well-defined habitats in montane regions that exist as fragments between which there is comparatively little communication.

The application of population dynamics models in landscape research will be limited by their inherent complexity. At the practical level it is very difficult to develop deterministic models for systems which may have one or more stochastic components. The introduction of stochasticity into a population model means that Monte Carlo simulation is almost always essential to investigate system functioning. The use of this type of simulation approach imposes considerable time constraints on the utility of the modelling system. This is probably the severest limitation on the potential use of deterministic population models in landscape ecology because migration is likely to be a significant factor determining the distribution of many species of organism. From the purely practical viewpoint it is not practical to develop deterministic population dynamic models for large numbers of species at the scale of the landscape even assuming that the underlying ecological processes determining the abundance of each of the modelled species can be quantified. Process-based population dynamic models of this sort are probably only practical where the organism has comparatively few and clearly understood relationships with its surrounding environment and associated biota.

4.4 The development of an integrated system for evaluating the ecological consequences of land-use change at the landscape scale

The rapid changes in the UK landscape during the last half of the twentieth century have emphasized the need for coherent and flexible systems for investigating land-use issues. The Natural Environment Research Council (NERC) and the Economic and Social Science Research Council (ESRC) recognized that advances in land-use research could only be made by adopting a multidisciplinary approach. Only by analysing the fundamental economic, physical and ecological processes that operate in landscapes and their interactions is it possible to predict the consequences of land-use change. Research under the NERC/ESRC Land Use Programme has endeavoured to provide a framework for integrating these disciplines and provide a generalized system for investigating the consequences of land-use change. A central theme of the ecological component of this research was the need to provide a holistic system capable of investigating all types of land use issue for as wide a range of ecological features as possible. Consideration of the ecological constitution of the landscape indicates that this is no easy task.

The simple change from a pastural to arable agricultural land use described above could result in the wholesale loss of more than 200 plant and animal species per square metre. It is clear that in seeking to investigate the ecological consequences of land use change at even the small scale, the ecologist is faced with a poorly structured problem of quantifying the effects of variation in a large number of land-use variables on a large number of species. The ecologist requires what is, in effect, a multiple response model with many driving or explanatory variables. Consideration of the range of landscape features in even the least diverse landscapes such as those dominated by arable agriculture, indicates that a multiple response model of this form would be even more complex when considered at the large scale. Whereas there are clearly many different potential approaches for investigating the ecological consequence of land-use change, there is no clear rationale for deciding on which procedures are most appropriate. As with any modelling exercise it would be logical to adopt a principle of parsimony. If a generalized system is required then it must satisfy a number of objectives. First, it should use the minimum number of numerical techniques, be they associative or deterministic, otherwise it is not generalized. Second, it should provide the maximum information on the responses of the largest number of species. Third, it should be capable of being used to investigate the largest number of land-use variables. Fourth, it should be statistically verifiable, validatable and robust. Devising a system that meets all of these requirements is by no means a simple problem.

One obvious method of rationalizing the problem of what to model is to limit the number of response variables, that is the types of organisms and communities considered in the modelling system. In the United Kingdom alone there are in excess of 25000 species of invertebrate, more than 2500 flowering plants, 250 birds and 50 mammal species and it is obviously not realistic to consider producing systems for investigating the effects of land-use change on all of these. Rather than exclude species on a piecemeal basis, taxonomic groups should be rejected or included on the basis of objective criteria. Definition of these criteria is not difficult since they are, in many ways, similar to those used in conservation. Taxa should only be included if they can be considered an essential component of an individual ecosystem or are responsible for determining the suitability of ecosystems for other groups of organisms. Flowering plants and bryophytes are the most obvious group in this respect and they are often considered paradigms of habitat in their own right. Taxa should be included only if there is also sufficient information on the basic ecology of the group at the family level. The great majority of the invertebrate taxa could probably be excluded on these grounds. Furthermore, if there is insufficient information available to allow the development of conceptual models both for describing the spatial distribution of a species and the potential effects of land-use change on these distributions, then the taxa should not be considered for inclusion. This may appear logical, but it is possible to produce models that satisfy the first objective of predicting the spatial distribution of a species, but which cannot be used as a component of a model for predicting the consequences of land-use change. Thus, species for which key habitat features determining spatial distribution are not directly relatable to land use should be excluded. Taxa should perhaps only be included if they also have a high public profile. Whereas some taxa may be important in determining the structure and functioning of individual ecosystems, there may be little point in including them if the group has little or no public appeal. Earthworms for instance, are important in determining nutrient cycling in many ecosystems but they have a low public profile.

Restricting the number of species considered reduces the problem of having a large number of response variables in the modelling system, but it does not help with the problem of having potentially very many explanatory variables. Rationalization of the processes involved in land-use change to a few key, driving variables would be an ideal solution, but any simplification must be ecologically realistic. One method of simplifying the number of potential land use change variables is to classify the effects that they have on the ecological structure of the landscape. If variables can be grouped hierarchically then it should be possible to model the effects of land-use change in terms of a few key variables at different levels in the hierarchy. Simplification of this type requires a classification of landscape in ecological terms.

The range of uses to which the land surface is put and the interest-groups associated with it has meant that there has been considerable interest in the development of systems for classifying and describing land. Where these descriptions are spatially referenced they provide a system by which the land surface can be mapped and the quality of the ecological components within it quantified. Even in a country as small as the United Kingdom, the number of land-cover and land-use classifications produced has been large and seems to be almost as great as the number of studies of land use undertaken.

The Nature Conservancy Council (NCC) developed a standardized system of biological survey, the Phase I methodology (NCC, 1990) which has been widely used to map wildlife habitats. The aim of this methodology was to produce a rapid recording system for seminatural vegetation and wildlife habitats over large areas of countryside. It aimed to provide an objective basis for determining which sites deserved consideration for conservation. The Phase I system is a hierarchical classification with basic habitats such as woodland subdivided into a range of subhabitats. Whereas the ecological information provided by Phase I survey is very detailed, the field survey and the storage and analysis of data are labour intensive, and large-scale mapping may be prohibitively expensive. More recently, satellites have provided imagery of ground cover that has been used to create land-cover classifications. Typically these classifications are derived from partitions of the radiation spectra reflected from the land surface. This type of information is increasingly used as a source of land cover information. The Institute of Terrestrial Ecology, Monkswood, produced a maximum likelihood classification of spectral reflectance data from the LANDSAT 5 Thematic Mapper at 25m resolution (Bunce *et al.*, 1992). Satellite imagery has the distinct advantage that cover is complete, relatively cheap and can be readily updated. The MLURI agricultural land capability classification (Bibby *et al.*, 1991) is a land classification that is directed at agricultural production synthesizing climatic, gradient, soils, erosion and wetness characteristics. The classes summarize the suitability of the land surface for agriculture activity and range from 'land capable of producing a wide variety of crops' to 'land of very limited agricultural value'. At a coarser scale of resolution the Institute of Terrestrial Ecology developed a classification of the land surface at the scale of the 1km national grid (Bunce *et al.*, 1981). This classification is based on physiographic features rather than the land cover itself and describes 32 basic classes. All 1km squares in the National Grid of the United Kingdom have been assigned to one of the 32 classes and the average cover of 128 habitat types in each class has been determined at three different times by field survey (Bunce *et al.*, 1992).

These land classifications can all be considered different formulations of the top level of an ecological-landscape hierarchy (Allen and Starr, 1982), comprising land cover/habitat, community and species. The land cover habitat level attempts to encapsulate all the structural variation such as whether or not the habitat is areal or linear. At the intermediate level are communities which are groups of species occurring together in the habitats, of which there may be many different types within each broad habitat grouping. The species form the lowest level in the hierarchy but are not necessarily exclusive to any one community type. These classifications make the assumption that variation between and within levels in the hierarchy is disjunct rather than continuous. There has been considerable debate about whether ecological variation in seminatural habitats is continuous or disjunct, with the consensus favouring the concept of a continuum of variation (Austin, 1990; Collins *et al.*, 1993). Although land classifications may not be the most appropriate method for describing ecological variation, classifications do form a convenient tool for modelling because land-use change can be rationalized into changes in the number and type of a finite number of landscape categories.

In the context of this ecological/landscape hierarchy, changes in land use can be classified as being of two kinds, either of type or of intensity. Changes in type arise where whole cover types, the highest level in the ecological hierarchy, are changed from one sort to another. Changes in intensity result in more subtle changes, lower down in the ecological hierarchy, at the level of the assemblage/community or the species composition within the individual community/assemblage. A good example of the first type of change is that resulting from a complete change in land use from one type to another as in the change from a pasture to an arable system. An example of the second change in land use would occur when an agriculturalist reduces the stocking rate of livestock on a pasture or when the intensity of use of fertilizers is reduced. Both of these changes in land use would be expected to have dramatic effects on the species composition of the land surface but they would not change the habitat type at the highest level in the hierarchy over the short term. This division into type and intensity is arbitrary because it is dependent on the scales of resolution of the levels in the hierarchy. It is also obvious that in some cases it may be possible for changes in intensity of land use to lead to wholesale changes in land cover. A very large reduction in stocking rate on moorland grassland for instance might be expected, over the course of time, to lead to the development of dwarf shrub communities. This is clearly an example of a change in intensity of land use having effects on the structure of the highest levels in the ecological hierarchy.

Reducing the number of species modelled and a rationalization of land use change into two basic types renders the problem of developing models for investigating the ecological consequences of land use change more simple.

4.5 The application of associative modelling approaches to investigating the ecological consequences of land-use change within the context of the ecological hierarchy

The essence of the problem is to extend knowledge of the distribution of species or communities in particular habitats to the likelihood of encountering species or species assemblages in space, using mapped habitat information from one or more of the habitat and land cover classifications. The effects of land-use change on distribution is then considered in terms of changes in the distribution of the mapped cover types and in the communities within these cover types. The taxa to be modelled can be divided into two categories. First, those species for which the habitat level of the ecological hierarchy is at sufficiently large a spatial scale to capture all aspects of habitat requirements. This group includes the plants and invertebrates. Second, those species for which components of several habitats are required to meet the overall habitat requirements of the organism which must move between the habitats to obtain all the resources it requires. This includes the larger more mobile vertebrates, particularly the birds and mammals. In both categories species may occur in more than one habitat type, but in the second group the presence of the species may require features of other habitats or features not specifically associated with a single habitat. The distinction between these two groups of species indicates the need for at least two associative modelling approaches for predicting species distributions and their responses to land use change.

4.5.1 SPECIES THAT UTILIZE ONE OR MORE HABITAT STATE EXCLUSIVELY

This is effectively a four-stage process whereby information on the occurrence of species within communities is linked to the occurrence of communities in habitats, the spatial distribution of which is known. A flow chart illustrating the processes is shown in Figure 4.1.

1. *Defining assemblages or communities* Much research in plant and animal community ecology has been concerned with identifying communities of organisms. In many cases the objectives of such research have been to identify possible factors determining the distribution of species assemblages. Most of these investigations have been based on the use of classification algorithms which provide objective methodologies for partitioning multivariate datasets. In the case of the plants, these approaches have been used to produce standardized descriptions of named and systematically defined vegetation types in the British Isles. The National Vegetation Classification (Rodwell, 1991) provides the most comprehensive information for plant species available. This was derived from a 15-year survey and analysis of vegetation in the British Isles initiated by the Nature Conservancy Council and undertaken by the University of Lancaster. It is comprehensive to the extent that it covered all natural, seminatural and major artificial terrestrial and aquatic habitats, but it excludes sown leys and communities where non-vascular plants are dominant (Rodwell, 1991). In excess of 35 000 samples were processed for this classification and it comprises the best data currently available on British plant communities. Equivalent classifications for invertebrate taxa have not been produced and are not

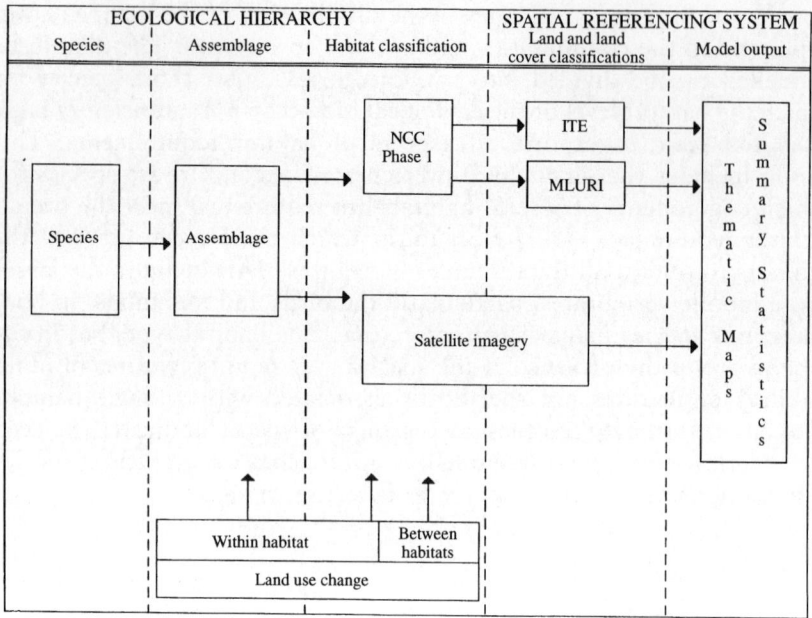

Figure 4.1 Flow diagram illustrating the use of an ecological hierarchy to investigate the effects of land use change on the distribution of species that utilize one or more habitat state exclusively.

available except for specific habitat types. Luff *et al.* (1991) and Rushton and Eyre (1992) produced classifications of ground beetle and spider assemblages for grasslands. Classifications for riparian invertebrate communities have been developed by Moss *et al.* (1987) and for aquatic coleoptera by Eyre *et al.* (1986).

2. *Quantifying the distribution of assemblages or communities within individual habitat types* In order to be able to predict the occurrence of individual species within habitats it is necessary to quantify the likelihood of encountering the assemblages or communities in which they are found within each habitat type. The most widely available information on the distribution of habitat information in the United Kingdom is that of the 25 land cover types derived from the Landsat satellite imagery or the cover types recognized within the NCC Phase I methodology. Habitat information derived from the satellite imagery is complete and potentially renewable at short time intervals; the collection and collation of the NCC Phase I information has, however, been less complete and has usually been undertaken at the county level. There have been few attempts to link the distribution

of habitat types with the community composition, although the former Nature Conservancy Council produced a contingency table listing the communities of the National Vegetation Classification likely to occur at up to the third level of the Phase I hierarchy (NCC, 1990). These data were only adjacency matrices (*sensu* Barnett, 1990) and gave no information on the relative abundance of each community within each habitat type. For both land cover classifications it was necessary to quantify these probabilities at greater resolution. The extent to which individual communities could occur within specific habitat types can be assessed from knowledge of the environmental conditions known to be favoured by each community and measures of these environmental features in specific areas where habitats were found. Sanderson *et al.* (1995) describe a method for doing this based on ordination procedures.

3. *Predicting the spatial distribution of individual species in the landscape* Mapping the distribution of individual species and assemblages is dependent on the use of spatially referenced habitat information. In the case of the satellite-imagery each parcel has a unique set of co-ordinates which can be used to position the parcel on a map. Thus, if the likelihood of encountering species or assemblages within each satellite class is known, a map of the probability of encountering the species or assemblage can be produced.

4. *Predicting the effects of land-use change* Adoption of categorical classifications for defining the ecological attributes within a catchment means that compartmental or matrix models can be used. In this case land-use change can be explained in terms of transitions between compartments within or between levels within the ecological hierarchy. Thus, changes in land use that lead to changes in land cover type are modelled by direct manipulation of the spatially referenced land cover/habitat data, whereas changes in intensity of land use result in changes in the relative contribution of different community types to that land cover/habitat type.

An application of this hierarchical approach to predicting the distribution of the plant species *Calluna vulgaris* (heather) is shown in Table 4.1. *C. vulgaris* is a small shrubby plant associated with habitats subjected to low intensity grazing on acid soils. The species is a major component of a number of recognized plant communities. In the catchment of the River Tyne there are at least 27 grass, moor heath and mire plant communities in 13 of which *C. vulgaris* may be found. The incidence of these plant communities in different habitat types depends on the habitat/land cover classification selected. In the case of the classified remote sensed imagery of the United Kingdom produced by the Institute of Terrestrial Ecology, these communities may be found in at least seven of the 25 classes recognized. The predicted distribution

Table 4.1 The incidence of *Calluna vulgaris* in the ecological hierarchy of plant communities and grassland heathland and mire habitats in the catchment of the River Tyne

Species	NVC community types		Habitat types[a]
	MG5	*Cynosurus cristatus–Centaurea nigra* grassland	Meadow–verge
	CG10	*Festuca ovina–Agrostis capillaris–Thymus praecox* grassland	
	U1	*Festuca ovina–Agrostis capillaris–Rumex acetosella* grassland	
	U2	*Deschampsia flexuosa* grassland	
	U4	*Festuca ovina–Agrostis capillaris–Galium saxatile* grassland	Grass moor
	U5	*Nardus stricta–Galium saxatile* grassland	
C. vulgaris	U16	*Luzula sylvatica–Vaccinium myrtillus* tall-herb	Grass heath
	H9	*C. vulgaris–Deschampsia flexuosa* heath	
	H10	*C. vulgaris–Erica cinerea* heath	Dwarf shrub heath
	H12	*C. vulgaris–Vaccinium myrtillus* heath	
	H18	*Vaccinium myrtillus–Deschampsia flexuosa* heath	
	M15	*Scirpus cespitosus–Erica tetralix* wet heath	Upland bog
	M16	*Erica tetralix–Sphagnum Compactum* wet heath	
	M18	*Erica tetralix–Sphagnum papillosum* mire	Lowland bog
	M19	*C. vulgaris–Eriophorum vaginatum* mire	
	M20	*Eriophorum vaginatum* mire	
	M25	*Molinia caerulea–Potentilla erecta* mire	

[a] Habitats not listed exclusively for each NVC community.

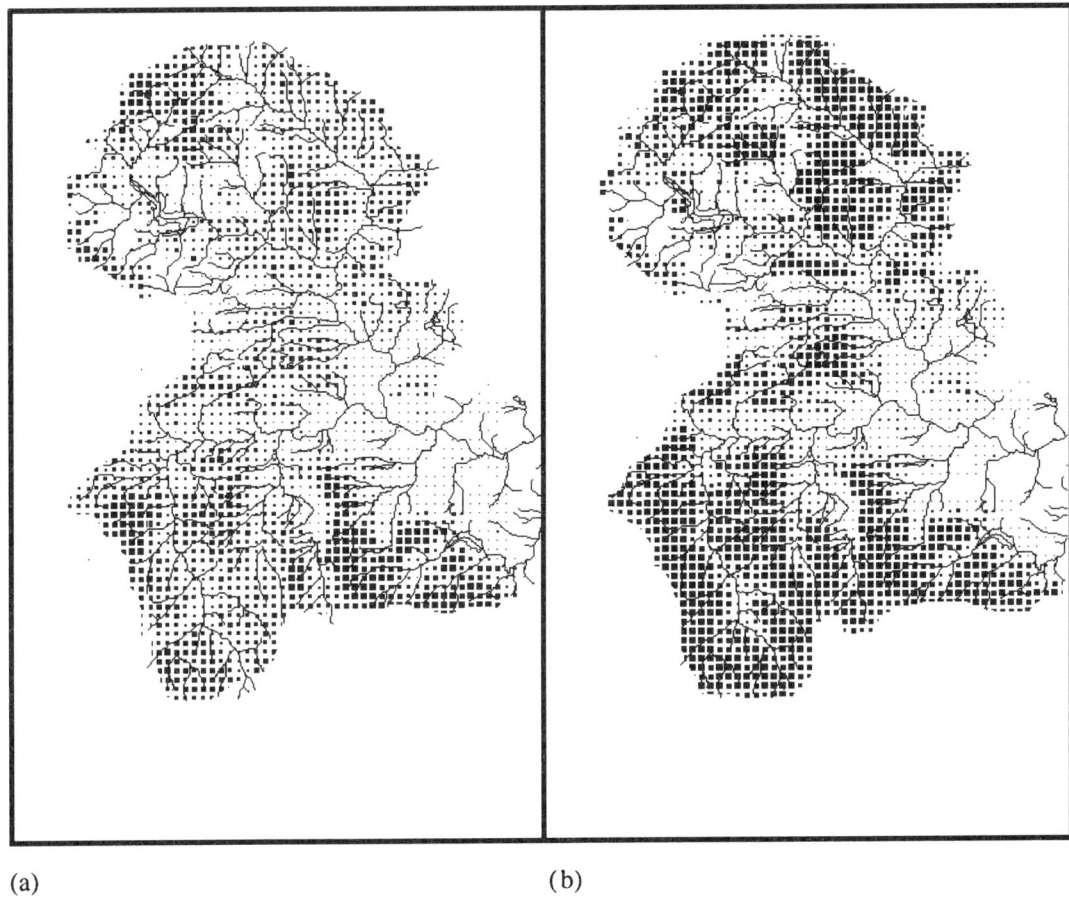

(a) (b)

Figure 4.2 Predicted distribution of *Calluna vulgaris* (a) before the imposition of land use change and (b) following a change from moorland grassland to heathland habitats. Grid square 1 km. Increased size of dot corresponds to an increased probability of encountering species.

of *C. vulgaris* in the catchment of the River Tyne using the ITE land cover classification is shown in Figure 4.2(a). The effects of changing grass moor habitats to dwarf shrub land cover as would result from deintensifying grazing is shown in Figure 4.2(b). The probability of encountering *C. vulgaris* increases across the whole catchment.

The utility of this approach to modelling land-use change has been investigated for plants and animals. Rushton (1992) and Rushton *et al.* (1994b) used the approach to analyse the effects of land-use change on the distribution of spiders and ground beetles in two river catchments in Northumberland, UK. For the spiders, the land cover data were derived from that used to describe the land classes of the Institute of Terrestrial Ecology Land Classification produced by Bunce et al. (1981). In the case of the ground beetle use was made of classified satellite imagery. Smith et al. (1992) and Cherrill et al. (1995) used the same approach with surveyed land cover and satellite imagery, respectively, to investigate the effects of land-use change on plants in two regions of northern England. Comparison of the predicted distributions of both plant and animal species using the two land cover data sets indicated that the satellite imagery provided finer detail in the mapped output.

Where validation exercises were carried out (three of the four studies) it was shown that the methodology was best at predicting common and non-ephemeral species. It was concluded that these limitations reflected the quality of the dataset as well as the underlying ecology of the species modelled. The statistical properties of this matrix modelling approach were investigated by Rushton (1992). These depend on errors within the original data and the extent to which these are propagated through the modelling system. The variance associated with a predicted probability of occurrence for an individual species is based on the expected propagation of errors through a series of multiplication and addition procedures that arise as a result of matrix multiplication. In addition there are several sources of error associated with the individual data matrices. The extent to which this methodology could be used to generate confidence limits for assessing the consequences of land-use change was investigated by Rushton (1992).

4.5.2 SPECIES THAT UTILIZE MORE THAN ONE HABITAT

For many bird and mammal species, the presence of one habitat type in the ecological hierarchy is not a good description of the suitability of an area for occupation by the species. Some species have requirements that require specific habitat and non-cover associated features. The golden plover for instance is found almost exclusively on heath moor but only on level ground and generally above 200 m altitude (Ratcliffe, 1978). Other species have more complicated requirements which include aspects of more than one habitat type, none providing all possible resource requirements. In these circumstances, the juxtaposition of habitat types in space, the landscape complementation (*sensu* Dunning et al., 1992) may be important in determining the overall suitability of an area. Large mammals, such as badgers, prefer sheltered, preferably sloping sites for setts and open pasture for foraging (Kruuk, 1987). Similarly, the golden plover will not nest in moorland habitats that are within 200 m of coniferous plantations. These apparently more complex habitat requirements arise in part because of the size and mobility of the animal relative to the scale of measurement of the habitat or cover-type, but they also reflect the ecologist's understanding of the behaviour of these species, since there is an abundance of general and detailed information on their habitat requirements.

Whilst there may be considerable information on the ecology of the birds and mammals, predicting their distribution in space, using hierarchical matrix models of the sort described above is not practical. Preliminary approaches using regression models linking land cover within individual 1km squares to observed distributions of birds have been proved to be coarse and inaccurate. These species can be thought of as interrogating the landscape; occupying it only if it is suitable and contains all necessary attributes for survival and reproduction. The most effective models for predicting the distribution of such species should be based on systems that emulate this behaviour. Developing computer algorithms that search landscape-information databases for suitable habitat features that may be at different levels in both a landscape and ecological hierarchy, and which also include complex spatial information on habitat complementation for more than a few individual species could be very complex. Behavioural models of this sort need to consider simple habitat variables, as well as 'meta-data', such as the spatial pattern of the habitats and their overall relationships with one another. Analysing large spatial environmental data in this way is inherently complex. The matrix approach for predicting species that are associated with one land cover type can be extended to include proximity data, provided facilities for collecting the data are available. These facilities are generally available in most geographical information systems (GIS). Rushton *et al.* (1995) utilized this methodology to investigate the effects of land use change on the distribution of 11 bird species. A flow diagram illustrating their modelling approach is shown in Figure 4.3. In this instance economic models were used to predict the responses of agriculture to the imposition of a price rise in nitrogenous fertilizers. The resulting land-use change caused changes at the highest level of the ecological hierarchy with a shift from arable to pastural agriculture, which had an effect on the distribution of wader species associated with these habitats. In this case the models predicted that there would be an increase in the likelihood of encountering the lapwing (*Vanellus vanellus*). Predicted distributions of red grouse (*Lagopus lagopus*) before and after the land cover change in the grass moor habitats of the catchment of the River Tyne are shown in Figure 4.4(a,b). Here the increased cover of *C. vulgaris* could be expected to have a marked effect on the distribution of red grouse since it is a major food resource for the bird. The modelling approach suggests that the distribution of red grouse would increase under this change in habitat.

98   Landscape ecology and land use

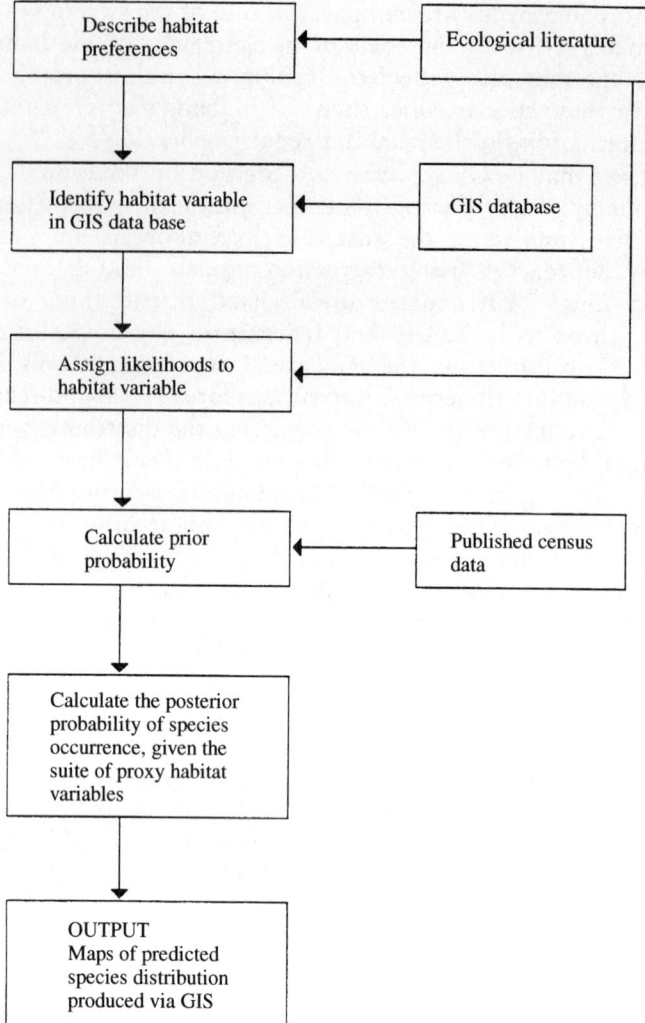

Figure 4.3 Flow chart illustrating the modelling system for predicting the distribution of species that utilize more than one habitat.

The development of models to investigate the effects of changes in intensity of land use within the context of a land cover–community–species hierarchy has been investigated for plants. Sanderson *et al.* (1995) developed an associative model for predicting the incidence of 534 species and 27 communities of the National Vegetation Classification, which can be used to predict the responses of species to changes in intensity of land use within habitats. In this model soil characteristics are used as an additional component of the ecological hierarchy to

The application of associative modelling approaches 99

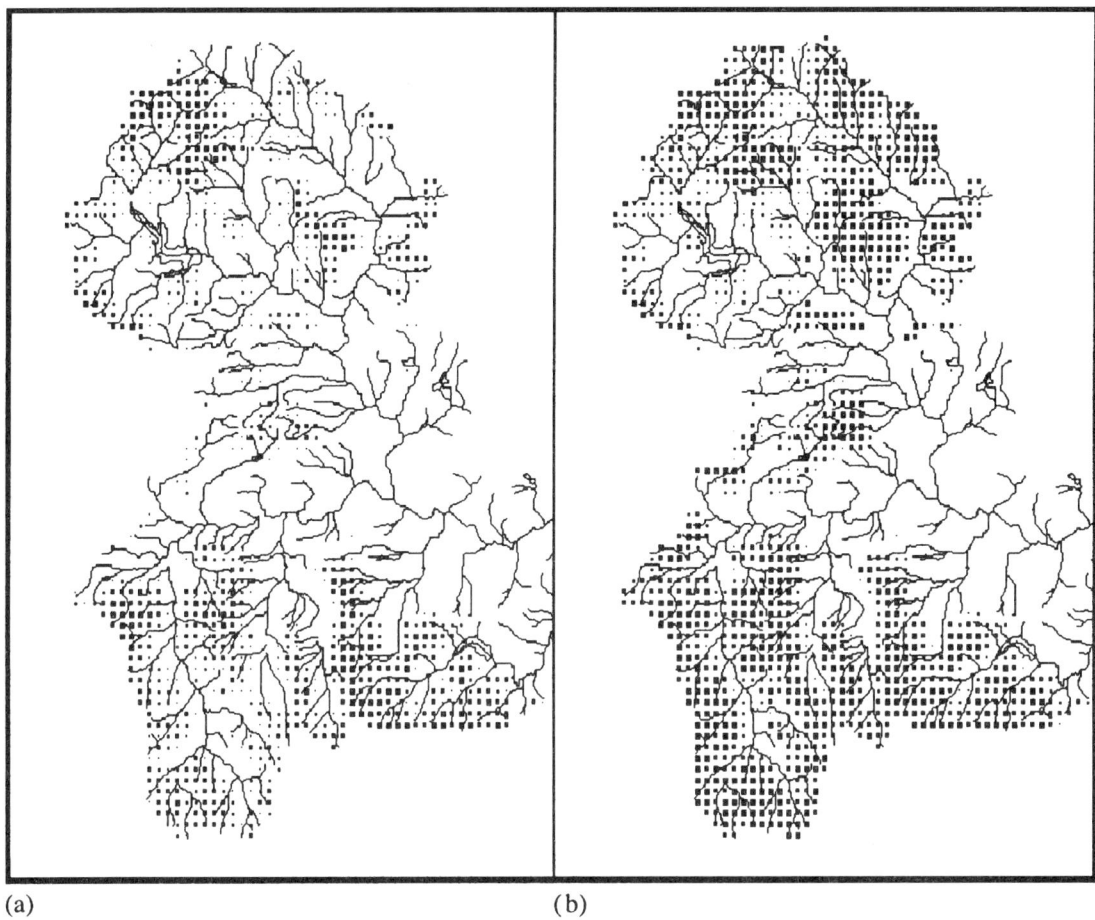

(a)   (b)

Figure 4.4 Predicted distribution of red grouse (*Lagopus lagopus*) (a) before the imposition of land-use change and (b) after the imposition of the land-use change described in Figure 4.2.

distinguish between the types of communities that can be found on different soil types. The modelling approach relies on the use of ordination to summarize data describing the environmental and management regimes under which different plant communities occur. The reduced dimensionality provided by the first two axes of the ordination then provides a two-dimensional space in which communities can be described. In effect the ordination represents a simplified, two-dimensional plot of habitat suitability for individual plant communities. The positions of real parcels of land in this space can be estimated on the basis of the ordination or component scores of the individual management and environmental variables that describe the land parcel. Changes in management result in changes in the relative position of points in the ordination space and hence allow the position of individual parcels of land to be tracked relative to those of the plant communities. A flow diagram of this modelling approach is shown in Figure 4.5. A graph showing the predicted occurrence of four communities of the National Vegetation Classifi-

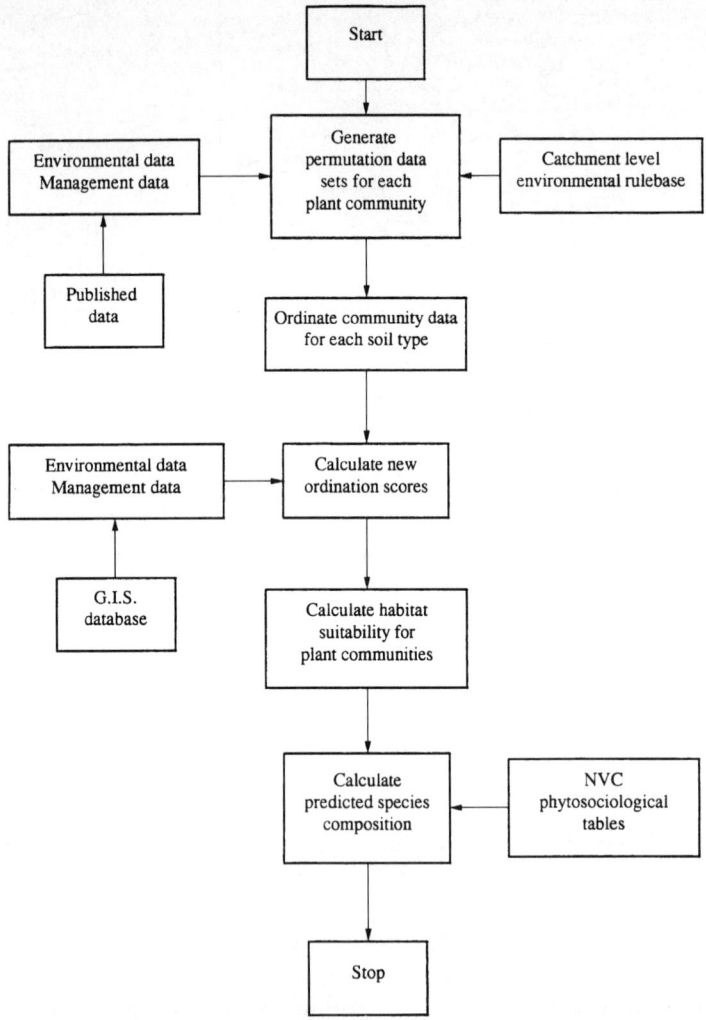

Figure 4.5 Flow chart illustrating the modelling system developed by Sanderson *et al.* (1995) to investigate the effects of vegetation management and environment on vegetation composition.

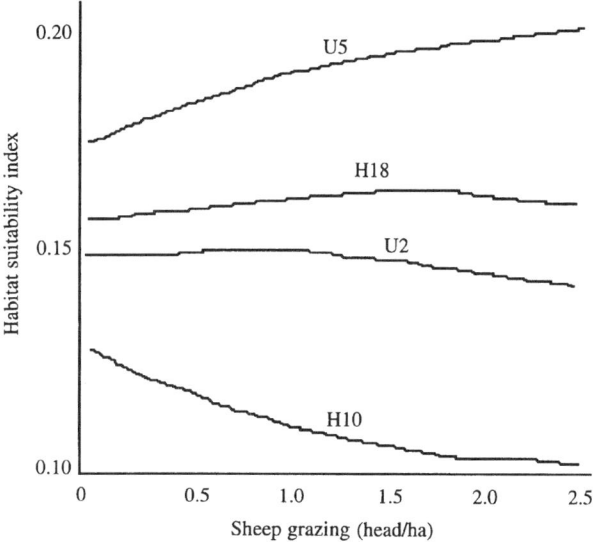

Figure 4.6 Predicted change in the likelihood of encountering four plant communities: *Nardus stricta/Galium saxatile,* Vaccinium myrtillus/*Deschampsia flexuosa, Deschampsia flexuosa, Calluna vulgaris/Erica cinerea* in relation to a change in grazing regime on a site at 300m altitude, 1000mm annual rainfall and a podsol soil. U5, H18, U2 and H10 are NVC community types (see Table 4.1).

cation in relation to changes in grazing regime, on a parcel of well-drained land on a podzolic soil at 400m altitude with mean annual rainfall of 1300mm is shown in Figure 4.6. The model predictions suggest that increased stocking rate will decrease the incidence of plant communities dominated by heather (*C. vulgaris*). This model has been used to investigate the effects of changes in nitrogenous fertilizer applications on lowland grasslands (Rushton *et al.* 1995) and the effects of reduced grazing and reduced fertilizer applications on upland heath and moorland communities respectively (Oglethorpe *et al.*, 1995).

4.6 The application of deterministic modelling approaches to investigating the ecological consequences of land-use change within an ecological hierarchy

Deterministic models link individual life history processes to the environmental factors determining them. These models can be used to investigate the consequences of land-use change if these processes can be linked to features in real landscapes. The number of processes that may be important in determining the spatial distribution of organisms are potentially very large. As with all the other modelling approaches considered, some degree of simplification is required before these methods become practical. Clearly, this modelling approach is only suitable for those species where the habitat requirements are simple and well known. Where species utilize only a few habitats of the range available in the landscape then all the life-history processes determining population abundance, with the exception of dispersal, can be considered to occur in these habitat types. This means that the top level in the ecological hierarchy can be divided into areas through which only dispersal takes place and areas where the remaining life-history processes occur. This means that all the processes that need to be modelled, except dispersal, are restricted to a few components of the top level in the ecological hierarchy and the relevant lower levels beneath them. Thus, the landscape can be divided into two and the modelling approach simplified into quantifying the effects of land-use change on dispersal processes in one set of habitats, and on life history processes in the remainder. A flow diagram illustrating the integration of population dynamics processes and dispersal through the landscape is shown in Figure 4.7.

The most suitable species for this modelling approach are mammals and birds. Species of high conservation interest such as the red squirrel are particularly appropriate because they occupy a restricted range of easily defined habitats, coniferous and deciduous woodland (Gurnell, 1987) and their spatial requirements are well known (van Apeldoorn *et al.*, 1994). Furthermore there is considerable background information on the factors affecting life-history processes of mortality and fecundity (Gurnell, 1987). A map of the predicted distribution of the red squirrel in the Tyne catchment under two land-use scenarios is shown in Figure 4.8. In this model habitat use was based on the animal utilizing woodland habitats with home ranges of 6.5 ha. The dynamics of populations were modelled assuming adult and juvenile mortalities of 40% and 75% and a maximum potential dispersal up to 10 km from site of birth.

A key advantage of such modelling approaches is that they can be built in a modular fashion. Thus they can be readily extended to include new features such as new forms of land use or additional information on the biological processes of a species. The value of such approaches has been recognized in the production of generalized modelling software such as that used by Lahaye *et al.* (1994) to investigate the dynamics of the spotted owl in California.

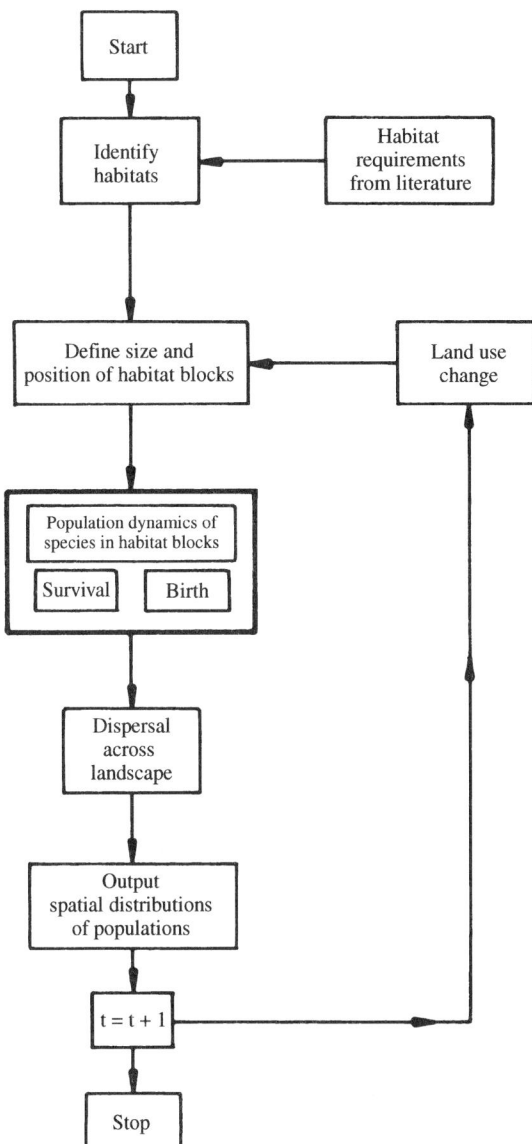

Figure 4.7 Flow chart illustrating a combined population dynamics GIS modelling approach for investigating the effects of land-use change on animal populations in landscapes.

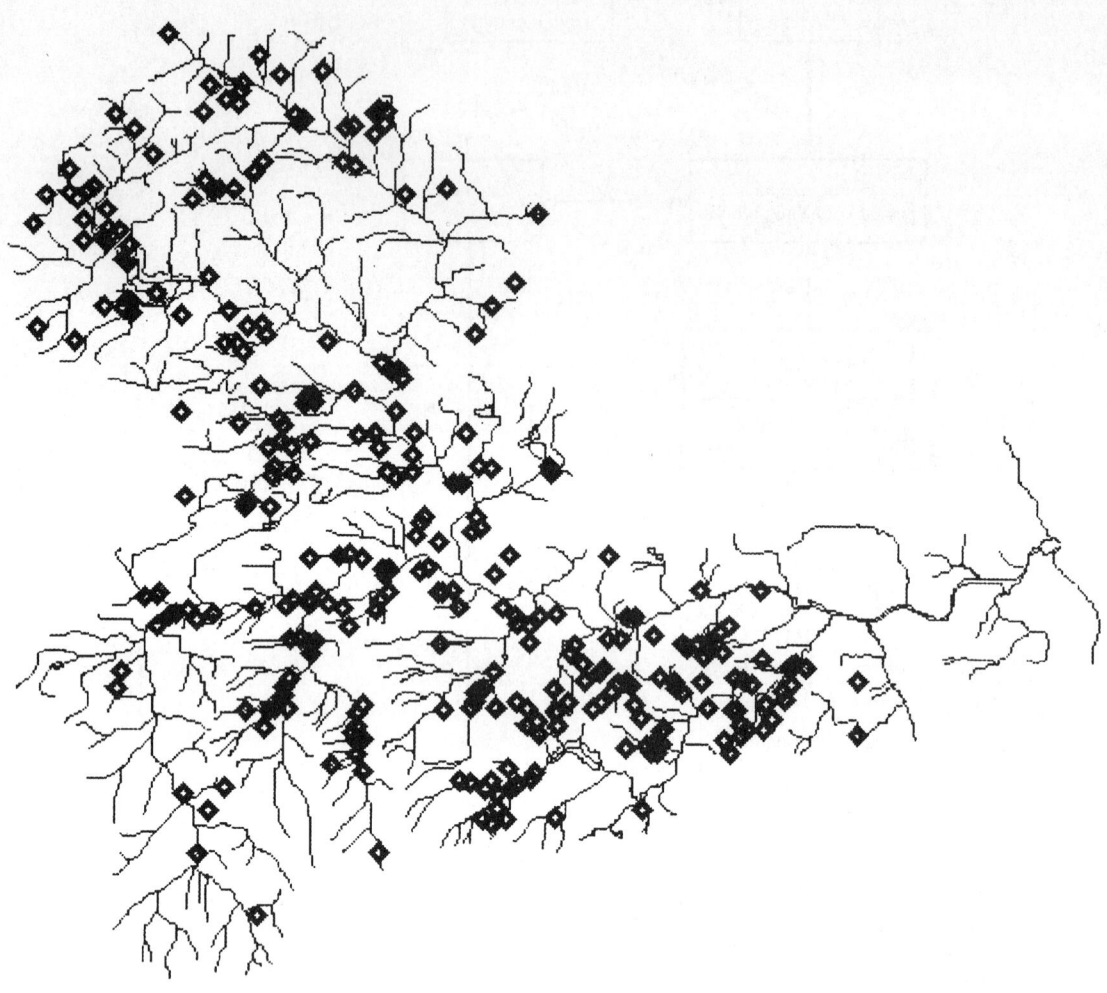

Figure 4.8 Predicted distribution of red squirrel in catchment of River Tyne as derived from a simple deterministic population dynamics model. Diamonds represent blocks of habitat of greater than 6 ha area in which populations persist given a reproduction output of six pups per pair per year, a juvenile mortality of 45% and an adult mortality of 40%.

Although the causes of land-use change are often driven by economics, it is obvious that the mechanisms and processes of land use change can be variable. Assessing the consequences of land use change for the organisms that share the land surface with the humans that use it is inevitably very complex. Humans, the other biota and the physico-chemical processes in landscapes operate at a range of scales. These lead to a hugely complex suite of potential interacting variables that inhibit attempts to predict the course and dynamics of land-use change. Recognition of the complexity of these processes has been a comparatively recent phenomenon and past study of land use has largely been piecemeal, with methodologies developed in response to individual land-use needs. Only with the development of landscape ecology as a subject has the need for an integrated holistic approach to land use and its effects on organisms become evident. It is perhaps in the nature of science that holistic approaches can only be developed from collation and synthesis of extensive and often apparently disparate knowledge. The holistic view is generated rather as a jigsaw is assembled, through a series of interlocking links, joined by thought. The analogy with the jigsaw perhaps fails to recognize the intellectual sifting necessary to perceive the whole picture. However, unlike landscape ecology most jigsaws have a finite number of parts with which to create the picture. Whatever the future of landscape ecology and land-use research there is no doubt that an integrated holistic approach can only be feasible if it takes account of the variation in scale and dynamics of the underlying processes determining landscape structure. From the viewpoint of the landscape modeller this will mean modelling within the context of hierarchies, the approach adopted within the NERC/ESRC Land Use Project.

4.7 Conclusion; whither land-use and landscape ecology?

References

Aitken, M., Anderson, D., Francis, B. and Hinde, J. (1989) *Statistical Modelling in GLIM*. Oxford Statistical Science Series 4 Oxford, 374 pp.

Allen, T.F.H. and Starr, T.B. (1982) *Hierarchy: Perspectives for Ecological Complexity*. University of Chicago Press, Chicago, 310 pp.

Aspinall, R. (1993) Use of geographical information systems for interpreting land-use policy and modelling effects of land-use change, in *Landscape Ecology and GIS* (eds R. Haines-Young and S.H. Cousins), Taylor and Francis, London, pp. 223–37.

Austin, M.P. (1990) Community theory and competition in vegetation, in *Perspectives in Plant Competition* (eds J.B. Grace and D. Tilman), Academic Press, New York, pp. 215–38.

Austin, M.P. and Gaywood, M.J. (1994) Current problems of environmental gradients and species response curves in relation to continuum theory. *Journal of Vegetation Science*, 5, 473–83.

Austin, M.P., Cunningham, R.B. and Fleming, P.M. (1984) New approaches to direct gradient analysis using environmental scalars and statistical curve-fitting procedures. *Vegetatio*, 55, 11–27.

Austin, M.P., Nicholls, A.O. and Margules, C.R. (1990) Measurement of the realised qualitative niche: an environmental study of five Eucalyptus species. *Ecological Monographs*, 60, 161–77.

Barnett, S. (1990) *Matrices, Methods and Applications*, Oxford University Press, Oxford, 450 pp.

Bibby, J.S., Douglas, H.A., Thomasson, A.J. and Robertson, J.S. (1991) *A Land Capability Classification for Agriculture*. Macaulay Land Use Research Institute, Aberdeen.

Bright, P.W., Mitchell, P. and Morris, P.A. (1994) Dormouse distribution: survey techniques, insular ecology and selection of sites for conservation. *Journal of Applied Ecology*, 31, 329–40.

Buckland, S.T. and Elston, D.A. (1993) Empirical models for the spatial distribution of wildlife. *Journal of Applied Ecology*, 30, 478–96.

Bunce, R.G.H., Barr, C.J. and Whittaker, H.A. (1981) An integrated system of land classification. *NERC Annual Report for 1980*, pp. 28–34.

Bunce, R.G.H., Carr, C.J. and Fuller, R.M. (1992) Integration of methods for detecting land use change, with special reference to countryside survey 1990, in *Land use Change: Cause and Consequences* (ed. M.C. Whitby), HMSO, London, pp. 69–79.

Chatfield, C. and Collins, A.J. (1980) *Introduction to Multivariate Analysis*, Collins, London, 246 pp.

Cherrill, A.J., McClean, C., Watson, P. et al. (1995) Predicting the distributions of plant species at the regional scale: a hierarchical matrix model. *Landscape Ecology*, 10, 197–207.

Collins, S.L., Glenn, S.M. and Roberts, D.W. (1993) The hierarchical continuum concept. *Journal of Vegetation Science*, 4, 149–57.

Dunning, J.B., Davidson, B.J. and Pulliam, H.R. (1992) Ecological processes that effect populations in complex landscapes. *Oikos*, 65, 169–75.

Eyre, M.D., Ball, S.G. and Foster, G.N. (1986) An initial classification of the habitats of the aquatic Coleoptera in north-east England. *Journal of Applied Ecology*, 23, 841–52.

Fitzgibbon, C.D. (1993) The distribution of grey squirrel dreys in farm woodland: the influence of wood area, isolation and management. *Journal of Applied Ecology*, 30, 736–42.

Forman, R.T.T. and Godron, M. (1986) *Landscape Ecology*, Wiley, New York, 619 pp.

Griffiths, G.H., Smith, J.M., Veitch, N. and Aspinall, R. (1993) The ecological interpretation of satellite imagery with special reference to bird habitats, in *Landscape Ecology and GIS* (eds R. Haines-Young and S.H. Cousins), Taylor and Francis, London, pp. 255–73.

Gurnell, J. (1987) *The Natural History of Squirrels*. Helm, London, 201 pp.

Haining, R. (1990) *Spatial Data Analysis in the Social and Environmental Sciences*, Cambridge University Press, Cambridge, 409 pp.

Hanski, I. and Thomas, C.D. (1994) Metapopulation dynamics and conservation: a spatially explicit model applied to butterflies. *Biological Conservation*, 68, 167–80.

Harrison, S. (1994) Metapopulations and conservation, in *Large Scale Ecology* and *Conservation Biology*, British Ecological Society Symposium Vol. 35 (eds P.J. Edwards, R.M. May and N.R. Webb), Blackwell, Oxford, 388 pp.

Hastie, T.J. and Tibshirani, R.J. (1990) *Generalised Additive Models*, Monographs on Statistics and Applied Probability no. 43. Chapman & Hall, London, 333 pp.

Hill, D., Rushton, S.P., Clark, N. *et al.* (1993) Shorebird communities on British estuaries: factors affecting community composition. *Journal of Applied Ecology*, 30, 220–34.

James, M. (1985) *Classification Algorithms*. Collins, London.

Jongman, R.H.G., Ter Braak, C.J.F. and van Tongeren, O.F.R. (1987) *Data Analysis in Community and Landscape Ecology*, Pudoc, Wageningen.

Kenkel, N.C. (1991) Spatial competition models for plant populations, in *Computer Assisted Vegetation Analysis* (eds E. Feoli and L. Orloci), Kluwer, Dordrecht, pp. 387–97.

Kruuk, H. (1987) *The Social Badger*, Oxford University Press, Oxford, 155 pp.

Lahaye, W.S., Gutierrez, R. and Akcakaya, H.R. (1994) Spotted owl metapopulation dynamics in Southern California. *Journal of Animal Ecology*, 63, 775–86.

Lenihan, J.M. (1993) Ecological response surfaces for North American boreal tree species and their use in forest classification. *Journal of Vegetation Science*, 4, 667–81.

Levins, R. (1969) Some demographic and genetic consequences of environmental heterogeneity for biological control. *Bulletin of the Entomological Society of America*, 15, 237–40.

Lotka, A.J. (1925) *Elements of Physical Biology*. Williams and Wilkins, Baltimore.

Luff, M.L., Eyre, M.D. and Rushton, S.P. (1991) Classification and ordination of habitats of ground beetles (Coleptera, Carabidae) in north-east England. *Journal of Biogeography*, 16, 121–30.

McCullagh, P. and Nelder, J.A. (1983) *Generalized Linear Models*. Monographs on Statistics and Probability, Chapman & Hall, London, 261 pp.

Moss, D., Furse, M.T., Wright, J.F. and Armitage, P.D. (1987). The prediction of the macro-invertebrate fauna of unpolluted running-water sites in Great Britain using environmental data. *Freshwater Biology*, 17, 41–52.

Nature Conservancy Council (NCC) (1990) *Handbook for Phase I Habitat Survey. A Technique for Environmental Audit*, NCC, Peterborough.

Oglethorpe, D.R., Sanderson, R.A. and O'Callaghan, J.R. (1995) The economic and ecological impact at the farm level of adopting Pennine Dales Environmentally Sensitive Area (ESA) management prescriptions. *Journal of Environmental Planning and Management*, 38, 125–37.

Okubo, A., Maine, P.K., Williamson, M. and Murray, J.D. (1989) On the spatial spread of the grey squirrel in Britain. *Proceedings of the Royal Society of London*, B, 238, 113–25.

Pacala, S. and Silander, J.A. (1985) Neighbourhood models of plant population dynamics 1. Single species populations of annuals. *American Naturalist*, 125, 385–411.

Perez-Trejo, F. (1993) Landscape response units: process-based self-organizing systems, in *Landscape Ecology and GIS* (eds R. Haines-Young and S.H. Cousins), Taylor and Francis, London, pp. 87–99.

Ratcliffe, D.A. (1978) Observations of the breeding of golden plover in Great Britain. *Bird Study*, **23**, 63–116.

Rodwell, J.S. (1991) *British Plant Communities*: Vol. 2 *Mires and Heaths*. Cambridge University Press, Cambridge.

Rushton, S.P. (1991) A discriminant analysis and logistic regression approach to the analysis of Walckenaeria habitat characteristics in grassland. *Bulletin of the British Arachnida Society*, **8**, 201–8.

Rushton, S.P. (1992) A preliminary model for investigating the ecological consequences of land use change within the framework of the ITE Land Classification, in *Land Use Change The Causes and Consequences* (ed. M.C. Whitby) HMSO, London, pp. 111–17.

Rushton, S.P. and Eyre, M.D. (1992) Grassland spider habitats in North-east England. *Journal of Biogeography*, **19**, 99–108.

Rushton, S.P. and Hassall, M. (1987) The effects of food quality on isopod population dynamics. *Functional Ecology*, **1**, 359–67.

Rushton, S.P., Luff, M.L. and Eyre, M.D. (1989) The effects of pasture improvement on the ground beetle and spider communities of upland grasslands. *Journal of Applied Ecology*, **26**, 489–503.

Rushton, S.P. Eyre, M.D. and Luff, M.L. (1990) *The effects of management on the occurrence of some Carabid species.* in *The Role of Ground Beetles in Ecological and Environmental Studies* (ed. N.E. Stork), Intercept, Andover, pp. 209–16.

Rushton, S.P., Luff, M.L. and Eyre, M.D. (1991) The habitat characteristics of grassland *Pterostichus* species. *Ecological Entomology*, **16**, 91–104.

Rushton, S.P., Hill, D.A. and Carter, S.P. (1994a) The abundance of riverine birds in relation to their habitats: a modelling approach. *Journal of Applied Ecology*, **31**, 313–28.

Rushton, S.P., Wadsworth, R.A., Cherrill, A.J. *et al.* (1994b) Modelling the consequences of land use change on the distribution of Carabidae, in *Carabid Beetles: Ecology and Evolution* (ed. K. Desender), Kluwer, Dordrecht, pp. 353–60.

Rushton, S.P., Cherrill, A.J., Tucker, K. and O'Callaghan, J.R. (1995). The ecological system of NELUP. *Journal of Environmental Planning and Management*, **38**, 35–52.

Sanderson, R.A., Rushton, S.P., Pickering, A.T. and Byrne, J.P. (1995) A preliminary method of predicting plant species distributions using the British National Vegetation Classification. *Journal of Environmental Management*, **45**, 265–88.

Smith, R.S., Rushton, S.P. and Wadsworth, R.A. (1992) Predicting vegetation change in an upland environment, in *Land Use Change The Causes and Consequences* (ed. M.C. Whitby), London, pp. 118–30.

Thomas, C.D. and Jones, T.M. (1993) Partial recovery of a skipper butterfly (*Hesperia comma*) from population refuges. *Journal of Animal Ecology*, **62**, 472–81.

Upton, G. and Fingleton, B. (1985) *Spatial Data Analysis by Example*, Vol. 1, *Point Pattern and Quantitative Data*, Wiley, Chichester, 410 pp.

Usher, M.B., Crawford, T.J. and Banwell, J.L. (1992) An American invasion of Great Britain: the case of the native and alien squirrel (*Sciurus*) species. *Biological Conservation*, **6**, 108–15.

van Appledoorn, R.C., Celada, C. and Nieuwenhuizen, W. (1994) Distribution and dynamics of the red squirrel (*Sciurus vulgaris* L.) in a landscape with fragmented habitat. *Landscape Ecology*, **9**, 227–35.

Verboom, J., Lankester, K. and Metz, J.A.J. (1991) Linking local and regional dynamics in stochastic metapopulation models. *Biological Journal of the Linnaean Society*, **42**, 39–45.

Volterra, V. (1926) Variations and fluctuations of the numbers of individuals in animal species living together, in *Animal Ecology* (ed. R.N. Chapman), McGraw-Hill, New York.

Wiens, J.A. (1989) *The Ecology of Bird Communities* Vol. 1. *Foundations and Patterns*, Cambridge University Press, Cambridge, 539 pp.

Wiens, J.A. (1992) What is landscape ecology really? Editorial comment. *Landscape Ecology*, **7**, 149–51.

Williams, B.K. (1983) Some observations on the use of discriminant analysis in ecology. *Ecology*, **64**, 1283–91.

Yee, T.W. and Mitchell, N.D. (1991) Generalised additive models in plant ecology. *Journal of Vegetation Science*, **2**, 587–603.

5 Decision support systems: the NELUP example

5.1 Introduction

When decision makers are formulating a policy, they need a large amount of information. They need information about the present situation on which to build a baseline. They also need to prepare forecasts of the effects of any proposed policy changes. The policy makers can then study the effects, and modify or justify the policy accordingly. Traditionally, forecasts of policy effects have been produced by a team of specialist researchers, who had a limited amount of information available in paper form. The predictions were often done by rule-of-thumb rather than precise analytical techniques. In recent decades, however, the information technology revolution has increased the amount of information available at the push of a button. It is now expected that information produced to explore the effects of a policy will be quantitative as well as qualitative. It is also expected that the information should be available quickly.

The different ways in which decisions may be reached and implemented, together with how data relevant to a decision may be incorporated into the processes leading up to a decision, are summarized in Figure 5.1. Within the disciplines of general management, the interactions between background information in the form of data, the techniques for analysing those data in order to reach a decision by selecting a particular course of action, and by whom that decision is implemented, may be placed in six fairly distinct subcategories.

Category 1 is reserved for an approach to decision making which is not as rare as we like to pretend. It is for those who do not wish to be 'confused by the facts', and consider that the best decisions are made using an instinctive sixth sense.

Category 2 is for decision makers, who do consider the data provided by the system they want to manage. Although the information may not even be held formally in a management information system, it may be

Category	Data provided by	Data analysed by	Alternatives generated by	Decision selection by	Decision implemented by	Approach to decision making
1	←——————————— Decision-maker ———————————→					Completely unsupported
2	↑ Information system ↓	←——————— Decision-maker ———————→				Information supported
3		← Model →	←——— Decision-maker ———→			Systematic analysis
4		← Model →		← Decision-maker →		Systematic analysis of alternatives
5		← Model →			← Decision-maker →	Systematic decision-making with over-ride
6		← Model →				Automated

Figure 5.1 Different approaches to decision making.

real enough, in that it has been accumulated through years of practical experience acquired by working in the system.

Management science postulates that decisions should be made through the exercise of intelligence in a methodical way by a paragon, who will consider all the facts that are available in a consistent way, and who will understand the processes involved in the problem. Such a decision maker would be expected to proceed in the following stepwise fashion:

- define the problem;
- generate alternative solutions;
- evaluate the alternatives;
- implement the best of the alternatives.

It is recognized of course, that not every decision maker has 'perfect knowledge' about all the factors at work in the system under consideration. However, there is likely to be better information about routine, recurring problems for which an optimum solution may be attainable, than about 'one-off' strategic problems, where the information is only sufficient, at most, for the generation of alternative courses of action. There is no single preferred method of reaching a decision about problems which are characterized by a high degree of uncertainty, but there is broad agreement that any general problem-solving process should generate alternatives, use quantitative techniques, whenever possible, and recognize that the outcome is more likely to be a 'satisficing solution' than an optimization.

Category 3 contains those sets of problems where a management science type model is used for the analysis of the data, which have been acquired as part of the description of the system under study. The use of a model makes the methods of analysis more formal and transparent. The accuracy of the analysis depends on how well the model represents the system and how representative the data are of the performance of the system. The decision maker will still be expected to generate alternative courses of action and to select and implement the best of the alternatives.

In category 4, the model assumes an additional function in that it generates the alternatives as well as analysing them.

In category 5, the model analyses the data, generates the alternatives and calculates the optimum strategy leaving the decision maker to carry out the operational changes that are required in order to implement the best of the alternatives. A linear-programming treatment of allocation problems is a well-known example of the category 5 approach.

Category 6 is exemplified by 'fly-by-wire' where an aeroplane in flight is entirely controlled by computers which collect the data, feed them into a simulation model of the flight control system, analyse the alternatives, select the optimum and make the necessary adjustments to the engines and flight controllers.

5.2 Modelling

Complex systems need to be simplified, so that they may be more easily comprehended. There is a long tradition of using simplified interpretations of real world situations, in order to render them more amenable to understanding and analysis, as well as providing a route to less costly experimentation. The iconic model is 'reality' at a reduced scale, as represented by a model aeroplane in a wind-tunnel, by plot experiments in plant breeding and by the models of buildings which architects sometimes use to demonstrate a design to a client. In an analogue model one physical property is used to represent another physical property, as by a needle moving over a gauge to represent speed, or differently coloured grid squares to represent population density. The third and most widely used form of representation is the mathematical model, in which sets of mathematical symbols and functional relationships are used to represent sets of physical situations. Mathematical models are precise, easily altered, amenable to computation and less expensive than either of the other two.

The process of mathematical modelling is illustrated diagrammatically in Figure 5.2. It begins with the observation and description of a 'real-world' problem, from which the modeller, by a process of abstraction, distils the essentials of the problem into some mathematical

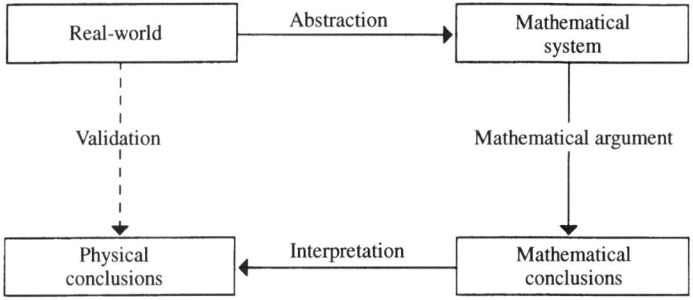

Figure 5.2 Formulating a mathematical model.

system, which represents it, but in a simplified way. The abstraction process is largely an art, which requires skill and experience, as well as considerable effort, imagination and expert opinion. A good model simplifies reality just enough to capture the principal modes of behaviour of the real system, without either over simplification to the point of misrepresentation or too much detail leading to a complicated analysis.

Following the abstraction, analysis based on mathematical argument is used to arrive at theorems or mathematical conclusions. The conclusions are then reinterpreted into physical deductions about the real world. This operation is, in part, the converse of the abstraction process and requires the same expert understanding of the 'real-world' system.

Simple models are simplifications of reality, different degrees of idealization of real-world systems are acceptable in different situations. The needs of those managing a system on a day-to-day basis are likely to be very different from those of the research scientist testing the underlying theory of a process. Although most real systems contain uncertainties, uncontrolled variables and unsteady states, it may be adequate for operational purposes to describe the steady-state performance using deterministic equations, in which it is assumed that the functional relationships are known with certainty. If, on the other hand, uncertainty cannot be ignored, abstraction based on stochastic models must be used to represent the functional relationships.

The physical conclusions, which are the outcome of the modelling process, must be checked with another set of data from the 'real-world' other than those used in the original abstraction of the mathematical system. The 'goodness of fit' between the test results and the predictions of the model is a measure of the validity of the model.

Success in modelling 'real-world' systems lies within a blend of extensive practical experience, applied theory, team work and good communication.

5.3 Decision support systems

Developments in management science have fostered approaches to decision making that place greater reliance on the use of data collected to measure the performance of systems and their component parts, on the generation of alternatives and on optimization. However, the acquisition of data is expensive, and the development of models requires theoretical sophistication tempered by operational experience. The effort and expense necessary to attain automated decision making can only be justified for a minority of systems, mainly those involving routine recurring problems for which there is a good database and where adherence to optimum operation is rewarded by an adequate pay-off. The pragmatic approach to the solution of structured strategic problems is modelling, based on the sources of information available, combined with human intelligence and experience.

Decision support systems (DSS) deal with poorly structured problems by helping decision makers confront them through direct interactions with data and analysis models. The underlying concept behind a DSS is that results should be available quickly without the need to consult experts. It is to be used directly by the actual policy makers rather than their specialist researchers. As early as 1981, Keen summarized the aim of DSS as

> designed to help improve the effectiveness and productivity of managers and professionals. They are interactive systems frequently used by individuals with little experience in computers and analytic methods. They support, rather than replace, judgement in that they do not automate the decision-making process nor impose a sequence of analysis on the user. A DSS is, in effect, a staff assistant to whom the manager delegates activities involving retrieval, computation, and reporting.
>
> Keen, 1981

The important components of a DSS are the software systems which provide the database management facility, the model management facility, and the dialogue management facility.

Within the field of computing, database management systems are an important subfield of research. Several types of data structures are used for database management. The simplest, for example a list of species names, hold data as simple flat files which are easy to create, edit and update. More complex data structures try to preserve the relationships between data.

In hierarchical structures, data are stored as nodes in an inverted tree structure, where the linkages between nodes represent 'parent–child' relationships. For example, an administrative area, such as a country, might be at the root of an inverted tree. The country might contain 50 counties. These counties form the next level in the tree. The links between the country node and the county nodes would represent the relationship 'is a county in'. Each county may be divided into several districts. The districts form the leaves of the tree. Storing data in such a structured way means that finding which children belong to which parents, or vice versa, is a reasonably straightforward operation. Queries beyond the parent–child relationships are not dealt with by this data structure. In the example given, all of the relationships are one-to-many, i.e. a district occurs only in one county, a county occurs only in one country. If the relationships are many-to-many, for example soils found in a particular district may occur in several other districts, this data structure leads to the same data being stored several times.

Network structures, as the name implies, store data as nodes in a network, where the links in the network between the nodes represent the relationships between the data. Considering the administrative area example again, but this time adding the extra relationship 'is a neighbour of' into the database, the tree structure becomes a network. This structure can handle more relationships, but all of the relationships must be thought out before the complex data structure is set-up otherwise a complicated dataset with many relationships may lead to a confusing structure.

Relational data structures are the most commonly used and are well documented (Codd, 1970, 1979). They have the advantage that they reflect the way in which we think about data and their relationships. Every dataset is held in a separate table, in which each item is stored in a row and the attributes are stored in columns. Relationships between the data are also stored as tables. The data structure is easy to set up and allows new relationships and data to be added easily and makes querying of the database relatively easy.

5.4 Geographical information systems

The obvious distinction between decision making in management science and in land use is the important significance of the spatial dimension. Spatial decision support systems (SDSS), draw on spatial models, which, in turn, depend on spatial databases for their information. Geographical Information Systems (GIS) may be considered as that part of database management, which provides both the technology and the software, for acquiring, storing, retrieving, manipulating and displaying spatial data. GIS largely display information in two-dimensional map form and concentrate on facilities for combining and analysing such data.

Just as with aspatial data, the structures used to represent spatial data digitally determine the ease with which the functions can be achieved. Two classes of spatial data structure, vector and raster are widely applied in commercially available GIS.

Vector data structures work on a data model of reality which deals with point, line and area data. This model fits conveniently with our perception of space, which is shaped by our understanding of maps and the way in which they represent certain features, depending on scale, as points (e.g. churches), lines (roads), and areas (woodland). The complexity of the vector data structure required to represent spatial data increases with the complexity of the manipulation and analysis which will be performed on them.

A simple 'spaghetti' data structure records a point as an (X,Y) co-ordinate along with a third value to indicate what the point is. Lines can be represented as strings of (X,Y) co-ordinates. The perimeter of an area can be stored in a spaghetti format as a line which forms a loop. However, spaghetti encoding stores none of the spatial relationships which we automatically interpret in maps. The simple data structure does not retain information saying which road is connected with which, or which roads form the boundary to a particular woodland area. More complicated spatial data structures are required to retain such information.

Fully topological data structures store the relationships necessary to allow data processing that determines which areas border which other areas, or which areas lie on either side of a line. All these topological data are required if maps are to be analysed in a vector GIS. In these, more complex, data structures, the data representing a line would include, not only the (X,Y) co-ordinates of the vertices of the line, but where the line starts and where it ends, together with the identities of the areas to the right and left of the line.

Vector data structures are compact and they provide an easily understood representation of traditional map elements. Only data for points and lines are stored, from which information about areas can be derived. Using a fully topological data structure the topology of a map can be completely described. Accurate graphics can be rendered from vector data structures, and database operations, such as update and retrieval, are achieved easily. However, a fully topological data structure is complex, because the relevant topological relationships must either be captured during digitizing, or calculated afterwards using complicated algorithms. Likewise, combining maps in overlay operations involves complicated algorithms requiring extra computer processor power. A further consideration, particularly for the SDSS environment, is that vector data structures result in irregularly shaped and sized areal units for which simulation models are more difficult to build.

Raster data structures are the most practical subset of a wider class of structures referred to as tessellation data structures. Tessellation data structures tile space with shapes. Each shape has a value representing the property of the space being represented digitally. The easiest shape is a square. Raster data structures cover space with squares or rectangular grids, the value given to a particular grid cell being determined by how much of the cell is covered by the property of the space recorded. The position in space of the cell is known because the origin and mesh size of the grid are known.

Raster data structures are very simple and the idea of grid cells in Cartesian space is easily understood. The resulting matrix structure is easily handled as is the overlaying and combining of maps. Spatial analysis, based on neighbourhoods around cells, is not difficult. Simulation models can be created more easily with raster than with vector structures.

Raster structures store data for every grid cell in space rather than merely boundaries to the areas making up that space. This can result in large volumes of data requiring storage, depending on the resolution of the grid mesh, but the problem can be alleviated by using compaction techniques. Choice of resolution not only affects data volume, but also controls what cartographic elements can be represented by the data structure. A grid resolution of 5m allows roads to be represented, but a grid resolution of 200m does not. The same resolution problem has an effect on graphical output from raster systems because the finer the resolution the more realistic a boundary will look.

The nature of the problems which a SDSS is intended to confront determines not only the data structures required but the GIS functionality that should be built into the DSS. Rhind (1990) classifies GIS operations by combining them with questions in the following way:

Location – 'What is at?';
Condition – 'Where does it occur?';
Trend – 'What has changed at?';
Routing – 'Which is the best way?';
Pattern – 'What is the pattern?';
Modelling – 'What if?'.

Location and condition are both database query problems and fall within the database management facility. The task of displaying the distribution of a spatial phenomenon, such as soil associations on a map, may appear trivial, but Crain and MacDonald (1984) point out that for the typical institutional GIS, these tasks will account for 50–100% of the applications of a GIS system during the first 4 years of its life. It is only beyond this initial phase of GIS development that more complicated analysis such as detecting trends and patterns will begin to be attempted. It will be even longer before modelling is likely to be employed in answering 'what if?' questions.

5.5 The NELUP spatial decision support system

The NERC/ESRC Land Use Programme (NELUP) was funded to investigate rural land use and the impacts of policy on the rural environment. In particular, NELUP was to construct a spatial decision support system that could help policy makers evaluate the consequences of policy change. The sciences of agricultural economics, hydrology and ecology were seen to be the prime factors in assessing the impact of change in the rural environment.

NELUP concentrates on the river catchment scale as its base unit. This scale was chosen primarily because this is the most convenient region for modelling hydrological processes. In addition, there is an increasing interest in management and planning at this scale, particularly with respect to pollution in rivers from pesticides and fertilizers.

The river catchments used in the development of the SDSS were the Tyne in the north-east and the Cam in the east of England. The Tyne catchment is approximately $3000\,km^2$ and includes moorland, pastures, arable and urban land covers. The Cam basin in East Anglia is approximately $800\,km^2$ and is predominantly an area of intensive arable farming.

The NELUP approach to a SDSS can be summed up in the following points:

- The interface should be user friendly;
- The database should be coherent and complete;
- The modelling capabilities should be powerful and accurate;
- A formal approach should not be imposed on the user;
- Display of information should be clear and easily interpretable;
- The system should have a rapid response.

The SDSS constructed by NELUP incorporates all of the above points. A database has been created that contains substantial amounts of information on the rural environment. A suite of models is available which uses the database to predict future states of the system. These predictions can be with either current policies, or user-implemented policies. The user can, for example, assess the effect of the GATT by lowering produce prices and subsidies.

The SDSS was developed with a raster rather than a vector database. This decision was influenced by the following.

- The core datasets were all supplied in raster format. To have converted these into vector format would have increased the complexity of the models and data manipulation, with no positive benefits.
- The main hydrological model works on a raster grid.
- Raster data can be manipulated quickly.
- When displaying output, the resolution of the model predictions is shown directly; converting to a polygon representation would make the output look misleadingly precise.

Many of the data which NELUP uses are supplied at 100m or better resolution. Unfortunately, however, the data were usually not sufficiently accurate to justify modelling at the base resolution. In addition, the computation times required for modelling at this resolution would be unacceptably long. Consequently, a resolution of 1km was selected, based on the National Grid of the Ordnance Survey.

All spatial data were converted from their base resolution into a 1km resolution at which the models were developed and validated.

5.6 The database

A substantial database is needed for NELUP. The data are needed for developing the models, running the models, and giving the SDSS user information about the region of interest. Efficient handling of these data, including error analysis and description, is essential.

A constraint placed on NELUP was that, although the SDSS should operate on a river catchment, it should be applicable to anywhere in Britain. In addition, it should be useable without undertaking much (or indeed any) fieldwork. Due to these constraints, NELUP has used nationally available datasets for the inputs to all the models. Although some field-work was done in the Tyne catchment, this was for model development and will not be necessary in order to apply the system to another catchment. Table 5.1 gives a summary of the data used in NELUP.

The data occupy many hundreds of megabytes of disk storage. Data for NELUP arrived in a number of file formats, and with errors, omissions and inaccuracies. For these reasons, a central database and structured data flow was developed as summarized in Figure 1.5.

A prime concern, when developing the database, was that it should

Table 5.1 Summary of the data used in NELUP

Data type	Variable	Source	Format
Terrain	Elevation	OS/IH	Raster map
	Gradient and aspect	Derived from elevation	Raster map
Rivers	River network	Institute of Hydrology	Vector map
	River flow	NRA	Hourly point time series (14 locations)
	River nitrate concentration	NRA	Monthly point time series
Meteorology	Temperature, sunshine and wind	Meteorological Office	Hourly point time series (16 locations)
	Rainfall records	NRA	Hourly point time series (64 locations)
Soils	Associations and classes	Soil survey and land resource centre	Raster map
Land classification	ITE land classification 1990	ITE Merlewood	Raster map
	Agricultural Land Capability (MLURI)	Derived from soil, climate and terrain data	Raster map
Land cover	ITE land cover map of Great Britain	ITE Monks Wood	Raster map
	NCC Phase-I Survey	In-house	Sample of 1 km squares in the Tyne catchment
		Northumberland National Park	1 km square summaries
Land use	Annual agricultural census	MAFF	Parish summaries
Farm management	Farm business survey	FBS North	Anonymous farm records
Flora	National vegetation classification	Lancaster University	Hierarchy of plant communities
	Plant species distributions	Various surveys	Raster map
Infrastructure	Road Atlas	Bartholomews	Vector map
	Administrative boundaries	Ordnance Survey	Vector map

Abbreviations: NRA, National Rivers Authority; ITE, Institute of Terrestrial Ecology; MLURI, Macauly Land Use Research Institute; NCC, Nature Conservancy Council; MAFF, Ministry of Agriculture, Fisheries and Food.

hold the definitive version of each dataset. When data arrive in NELUP, they are loaded into the database and checks are performed which test the reasonableness and reliability of the data. If necessary, the data are modified to remove obviously incorrect information.

Most of the data used in the DSS and the model systems are derived from one or more base datasets. Derivation of datasets from these core data was done either by the database software, or by in-house programs. The derived data were loaded into the database for general NELUP use. Three primary software packages were used in the construction and management of the database:

ORACLE

This is a relational database management system (RDBMS) (Oracle, 1989). It uses an extension of the internationally standardized SQL (structured query language) for data access and manipulation. The Oracle RDBMS is primarily used to store aspatial and time series data.

GRASS

This is a public domain raster GIS (Westervelt *et al.*, 1990). It is used as the primary store for all raster maps. In addition, it provides powerful means of data analysis and manipulation.

ARCINFO

This is a popular, primarily vector, GIS (ESRI, 1991). It is used in NELUP for digitizing, analysis of field surveys, and the storing of other vector datasets.

5.7 The models

The core of the NELUP approach is the integration and interfacing of three sets of models and associated data. It is the combination of these models that gives the DSS the ability to analyse policy options. The models are described in Chapters 2, 3 and 4.

As mentioned earlier, the DSS has been implemented on a 1km raster grid. The models have therefore been developed and validated at this resolution, taking into account errors associated with the base data. Although the purpose of the system is to make predictions about future states of the rural environment, the models can only be validated using analyses based on past changes in land-use policy.

5.7.1 AGRICULTURAL ECONOMICS

The two models available in agricultural economics are a regional model, and a farm-level model. Both models use linear programs (LP) to maximize profit by manipulating land capability, stocking rates, chemical application rates, types of crops and intensities of management.

The regional model treats the study region as one large farm. Initial conditions needed for the model are derived from the MAFF June Farm Census. These data detail, on a parish basis, land use, stocking rates, labour usage and farm types. NELUP has used areal interpolation methods to disaggregate these data to the MLURI land capability classes. Farm management practices needed for the model are derived from information gathered by The Farm Business Survey. These data provide time series information for an anonymous sample of farm enterprises. Crop modelling requires information about yields as a function of land capability class and fertilizer application rates. These yield parameters have been calculated using the EPIC software. The regional model produces land use changes which are converted into cover changes in the land cover map. The model was validated by initializing it with the MAFF land-use data for 1981, running the model for 7 years, and comparing the model output for each year with the MAFF land-use data.

The farm-level model gives a more precise picture of the activities on representative farms by incorporating greater detail of farm management practices into the LP. Like the regional model, the FBS and EPIC are used to calculate model parameters. Data from individual FBS farms have also been used to validate the model.

5.7.2 ECOLOGY

Two types of ecological model are used. The first is an associative matrix model that predicts the distribution of plant communities, plant species, and invertebrates. This model has been developed from the combination of published data, ecological surveys and the NELUP spatial database. The model outputs the likelihood of encountering a species within a grid-square, given the land cover.

The second type of ecological models are more process oriented. Two models fit into this category: a system for predicting the distribution of breeding bird habitats, and a multivariate prediction of the distribution of plant communities. The bird habitat model uses the spatial database to generate a Bayesian likelihood for each grid-square. Plant communities are predicted by combining the results of an ordination analysis, the spatial database, and the land use output of the economics models, to produce a habitat suitability index for each land parcel. These models have been validated against sample data for the Tyne catchment.

5.7.3 HYDROLOGY

Two hydrology models may be accessed through the DSS. One is the macro-model NUARNO, which works at a subcatchment resolution with a one-hour timestep. The parameters modelled are overland, ground-water and channel flows, and components of evapotranspiration. Outputs are available either as time series at the subcatchment outflows, or in the form of subcatchment mass balance diagrams. The NUARNO takes approximately 6 min per year to run for the whole catchment.

The second, SHETRAN, is a derivative of the Systéme Hydrologique Europèen (SHE). SHETRAN is a detailed, physically based, spatially distributed, modelling system that utilizes more than 60 hydrological parameters. Modelling is performed on a subcatchment basis, at a 1 km resolution, with an approximately one hour timestep. Over 50 output variables are available from SHETRAN, including details of ground, surface and channel flow, evapotranspiration and contaminant transport. These variables may be available as point time-series, map or in the form of subcatchment mass balance diagrams. Although SHETRAN can produce precise information at a high resolution, it is time consuming to run; a one-year simulation for the whole Tyne catchment takes about 48 hours.

The hydrological models use the ITE land cover map to distribute hydrological parameters such as evapotranspiration, interception and plant uptake. In addition, they use information from the regional economics model to distribute chemical applications and cropping intensities. Both models have been validated against historical records.

5.8 SDSS interface design

For the policy maker, the SDSS interface is the only access point to the database and models. The quantity of data is large, and the models complex, so a sensible interface design was an important step towards making the system user-friendly. If access to the data was provided via a few long lists of options, the interface would become cluttered and it would be difficult for the user to find relevant information. Likewise, if the interface was structured with a deeply nested menu hierarchy, it would take an excessive number of selections to get to the data, causing the user to become frustrated. The compromise reached was to group the options into a number of menus, each of which provides access to more detailed options. This means that the user can easily find the data they are looking for, with only a few selections needed.

The data have been grouped into physical, socioeconomic, agricultural economics, ecology and hydrology. Each of these menus is for easy communication with the system, without giving the impression of working within a rigid structure. There are also menus to access the model interfaces and utility functions.

Another major factor influencing the user-friendliness of the system is the graphics used to display data. The most important aspect of the graphics is that they must be consistent and readily interpretable. If they are too complicated, the user may become confused. If they are too simple, the user may get irritated. This again required

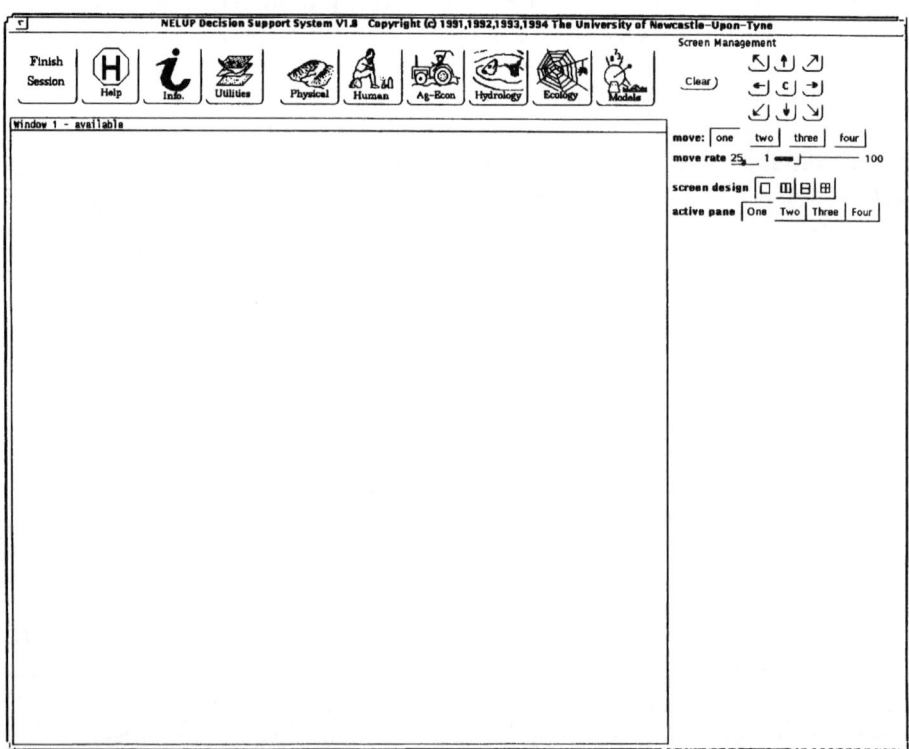

Figure 5.3 Start up screen.

a compromise, together with the sensible application of cartographic theories. For instance, colour can both make a display look more interesting, and also communicate information additional to that in a monotone image. However, over-use of colour, or use of inappropriate colours, can create a confusing or misleading display.

By the very nature of a spatially distributed system, most information displayed will be in the form of maps. As mentioned earlier, a raster data model has been adopted for the system. The maps displayed on the SDSS use a wire-frame grid, where the cells represent a 1km square. These wire-frame maps can be displayed as a standard two-dimensional view or can be draped over the elevation map to display a surface. There are three primary methods for displaying data in a grid cell, corresponding to three types of data. For displaying categorical values each square is filled with one of N colours, where N is the number of categories in the map. The second method uses a colour scale (or ramp) to display continuous data; the intensity or shade of the colour increases with the value of the attribute. The third type of map fills the grid-squares according to a probability or percentage value for specific attributes. For example, a square that is half full would represent 50%, whereas a full square would represent 100%. Other types of graphical output include bar-charts, X–Y plots and histograms. In addition, specific displays have been developed for some data, where these require different formats to ensure clarity for the user. Where appropriate,

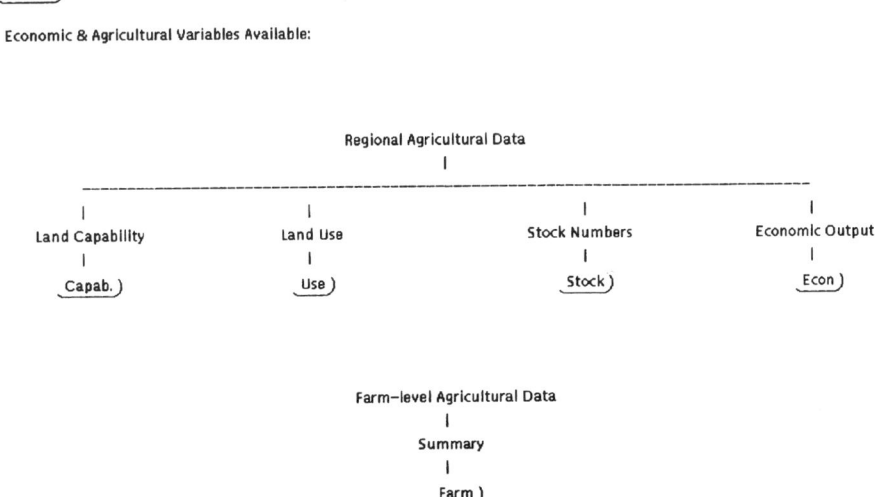

Figure 5.4 Agricultural economics menu tree.

a pop-up key providing more information is available for a graphical image.

When creating a display, the user has a number of options from which to choose. These options fall into two categories: specifying the data to display; specifying the appearance of the display. The options available to specify the data to display vary between one dataset and another, but retain a consistent layout. Options specifying the appearance of a display are the same regardless of the data being displayed.

5.9 Using the SDSS

An outline of the DSS can be gained from examining Figures 5.3 and 5.4. The figures give a general impression of the system, but do not give a complete resume of its capabilities.

Figure 5.3 shows the screen when the system is started up. The screen has four regions: the row of buttons along the top; the drawing window in the centre; the screen control options in the top-right; the blank area on the right side.

The row of buttons along the top of the screen are means of access to the database, models, help and information. Most of these buttons cause a menu tree to be displayed, as described later. The *Help* and *Info* buttons cause a pop-up window to be created containing context-sensitive help or information. The distinction between the two is that help describes the currently available options, whereas information gives a background to the data sources and the meaning of the generated output. The information system also describes the reliability of data and models.

The drawing window in the centre of the screen (presently labelled *Window 1: available*), is used for displaying graphical output. The menu tree created by the buttons described above is displayed over this region of the screen.

The *Screen Control* buttons, in the top-right corner, control the appearance of the drawing window. There are a total of four drawing 'canvases' into which graphical output can be stored. Either one, two or all four of the canvases can be displayed on the screen at any one time. Which canvas to use and the layout of the canvases is controlled from the panel.

The right side of the screen (presently blank) is used to display options needed for creating graphical images and controls.

Figure 5.4 shows the agricultural economics menu tree, created by selecting the *Ag-Econ* button from the top row, and is displayed over the top of the drawing window. The tree structure is used to imply the structure of the data, without imposing it on the user. Other menus are similar in design, but obviously offer different options.

A sample of the outputs and the way they are presented may be understood more clearly by consulting the World Wide Web.

In the spirit of the World Wide Web there is an interactive tour of the NELUP DSS located on the CLUWRR WWW server. The URL is http://www.cluwrr.ncl.ac.uk/nelup/nelup.html.

The tour begins with a brief description of the objectives of NELUP as an introduction to a demonstration of how a Graphical User Interface facilitates the communication of spatial information in a spatial way. The visitor may interact with the Decision Support System by clicking the mouse on images and hyperlinks in order to get a feel for the types of display and interaction that are possible between a user and the system. Samples of outputs from the DSS, showing both data and the results of model simulations, are displayed.

In addition, there are hyper-text links to WWW sites of various third party organizations whose products were used in the development of the NELUP DSS.

5.10 Building the DSS

In building an interface such as the SDSS, it is common to use the macro-language of a GIS. Macro-languages often include powerful numerical routines and can produce detailed graphics, within a short development time. The disadvantages, however, are that the programmer is generally limited by the capabilities of the package. For the NELUP DSS, it was decided that the interface would be written in-house. It was also decided that, due to the quantity of data and the size of models, the system would be developed on Sun workstations. The popular X11 protocol (Johnson and Reichard, 1992) which is now a widely used standard that is supported on all UNIX systems, was chosen for graphics display. The XView library (Heller, 1991) was chosen to implement the graphical user interface (GUI). XView is an OPENLOOK compliant GUI library, written by Sun Microsystems, that uses the X11 protocol to communicate with a windows server. The DSS interface, is completely portable to any UNIX workstation.

The software design methodology is a mixture of object oriented and modular (Lamb, 1988). The X11 library works in an object-oriented manner, trapping events and then calling programmer-supplier modules. The code that interacts with the GUI conforms to this design strategy. The code which provides the modelling and graphical output uses a modular design. Due to the very nature of research, it is likely that certain features of the DSS may need to be created, modified, updated or removed in the future. Therefore, the ability to alter certain sections of the system, without affecting others, was a core design feature.

Three software libraries have been written as part of the DSS. The first library implements a number of abstract data-types and utility functions. The data-types include raster maps, arrays, vectors, points and time series. This library is used extensively in all sections of the DSS.

The second library provides an interface to the database. A series of set-up files are used to define the location and format of a data file. The file access functions from the array library are used to read and interpret these files. A module in the DSS may request data without having to know anything about the location or format of the disk file. The library also implements a file buffering system to reduce disk accesses.

The final graphics library manages all aspects of the DSS graphics display window. Algorithms for creating the images are coded here, and are called from the display modules.

The economics and hydrology models are implemented separately from the DSS. The economics models are solved by use of the SAS statistical package's LP module. When an economics model is run, the DSS interface modifies the LP files and then calls SAS to solve the program. When SAS is finished, the DSS interprets the output and stores new land-cover maps in the database. In contrast, the hydrology models are both implemented as stand-alone FORTRAN programs. SHETRAN, due to the length of time it takes to run, is implemented in batch mode. The input files and options are set up on the DSS interface which then calls the appropriate commands to put it into the queue. When SHETRAN has finished executing, the user must tell the DSS to decode the SHETRAN output files and store them in the database. The NUARNO model is run interactively and the DSS waits for execution to finish before continuing.

The ecology models are all hard-coded into the DSS. The matrices of the associative model are precalculated and stored in the database. The parameters for the other models are also stored in the database and accessed on-line.

5.11 Assessing usability: workshops

Over a one year period (1992/93), a number of workshops were conducted with potential end-users of the DSS. The workshops were designed so that participants would gain an understanding of the work of NELUP, and also that NELUP could get user-feedback during the development phase of the research.

The workshop participants were selected so as to represent a broad range of interests relating to land use. In addition, a broad range of technical skills was represented, from GIS specialists, to those with no computer knowledge. This range of skills and interests enabled NELUP to assess better the utility of the SDSS.

The workshops, typically, took a full day. They comprised a description of NELUP, a concise demonstration of the SDSS, a hands-on session with the system, and a period of critical appraisal and discussion.

The hands-on session involved working through one of a number of prepared case studies. These briefly outlined a possible policy (e.g. proposals for CAP reform, regional afforestation) and then gave the steps needed to appraise the policy using the SDSS. The participants were free to explore the SDSS and work through the case study.

All workshop participants were extremely impressed with the system. The ease of access to complex models and the extensive database were seen as a major step forward; especially for those who had struggled with the complexity of standard GIS tools. The structure of the interface and grouping of data into the menus was seen as a good aid to understanding. Participants found the graphical display of data easy to interpret and the summary diagrams extremely useful. Some minor modifications to the interface were suggested, most of which have subsequently been implemented.

Less positive comments about the system mainly concerned the scale and resolution at which the system operates. For example, the needs of local bodies are for detailed information within their boundaries, while national bodies require information for the whole country. The river boards, not surprisingly, were satisfied with the catchment-scale. Many of the participants came from organizations which either possessed, or were in the process of developing, in-house GIS expertise. Comments were made that a system, with modelling capabilities and a comprehensive database, would be a worthwhile investment.

When using the DSS, it was surprising how few people questioned the data being put up on the screen. If the display looks impressive, then most users were prepared to accept that it was accurate. This is particularly worrying when the data are outputs from a model, which is, by definition, an approximation of the system. Other users (mainly those with GIS expertise) were more critical of the displays and were keen to understand the associated errors.

5.12 Conclusions

NELUP has created a spatial decision support system which can be used both for the formulation and monitoring of regional policies with particular reference to rural areas. It is particularly appropriate for catchment planning. Although the approaches taken by the three modelling groups are not necessarily novel, the combination of these models to provide a coherent system is a unique achievement. The principle of making specialist information available to non-specialists has helped to break down barriers on information flows between disciplines and to improve the dissemination of technical knowledge.

The major problem encountered in building the system, in common with any GIS project, was the gathering and management of data. The management of data has been made much easier and less problematic by the use of a well-structured, centralized database. Access to a database which gives a comprehensive picture of the region being studied is of enormous value in itself. Population census data, for example, were not used by any of the NELUP models and were thus not available on the system. At every workshop, however, participants requested the data be included as it provides useful background information (it has subsequently been made available).

The interface has a significant impact on the user's perception of the system. It is important that the graphical output of a DSS displays data in ways with which the user can relate and easily understand. Complicated images, although displaying more information, can be difficult to interpret. The response time of the interface can greatly influence the user's empathy with the system.

NELUP has shown that a DSS which provides access to a coherent database and suite of models can provide policy makers with a powerful analysis tool.

References

Codd, E.F. (1970) A relational model of data for large shared data banks. *Communications of the Association for Computing Machinery*, 13, 377–87.

Codd, E.F. (1979) Extending the database relational model to capture more meaning. *Association of Computing Machinery Transactions on Database Systems*, 4, 397–434.

Crain, I.K. and MacDonald, C.L. (1984) From land inventory to land management. *Cartographica*, 21, 40–6.

ESRI (1991) *ARC/INFO User's Guide, Version 6.0*. Environmental Systems Research Institute, Redlands, California.

Heller, D. (1991) *XView Programming Manual*, 3rd edn. O'Reilly and Associates Inc., Sebastopol, California.

Johnson, E.F. and Reichard, K. (1992) *X Window Applications Programming*. MIS Press, New York.

Keen, P.G.W. (1981) Value analysis: justifying decision support systems. *Management Information Systems Quarterly*, 5, 1–16.

Lamb, D.A. (1988) *Software Engineering.* Prentice Hall, Englewood Cliffs, New Jersey.

Oracle (1989) *Oracle Database Administrators Guide: oracle rdbms version 6.0.* Oracle Corporation, California.

Rhind, D.W. (1990) Global databases and GIS, in *The Association for Geographic Information Yearbook 1989* (eds M.J. Foster and P.J. Shand), Taylor and Francis and Miles Arnold, London, pp. 85–91.

Westervelt, J.M., Shapiro, M., Goran, W. and Gerdes, D. (1990) *Geographic Resources Analysis Research System. Version 4.0 User's Reference Manual.* US Army Corps of Engineers Research Laboratories, Champaign, Illinois.

6 Deintensification of agriculture: an evaluation of two land-use strategies

6.1 Introduction

The early 1990s have been a period of rapid change in UK agriculture. There has been a move away from the strategy of encouraging production, through the use of high-input high-output farming techniques, towards lower intensity methods of farming. These changes have been driven by both environmental and economic forces. It has been recognized that intensive agricultural techniques have caused major losses of wildlife habitats and species, and food surpluses and the massive subsidies necessary to sustain the system have become politically unacceptable (Whitby and Lowe, 1994).

There have been significant policy changes in an attempt to address these problems, both in the context of UK agriculture, and in the European Union (EU) as a whole. The EU enacted a series of radical reforms of the Common Agricultural Policy (CAP) in 1992, the so-called 'MacSharry Reforms'. These reforms aimed to stabilize farm incomes and reduce overproduction through measures such as compulsory set-aside and the imposition of limits on livestock intensities. As a separate measure, the UK Government, through the Ministry of Agriculture, Fisheries and Food (MAFF), has sought to promote the use of less-intensive farming techniques via a number of initiatives, in particular the introduction of Environmentally Sensitive Areas (ESAs), Nitrate Sensitive Areas (NSAs), and Countryside Stewardship Scheme (the latter in partnership with the Countryside Commission). Although only a relatively small area of agricultural land is covered by these schemes (c. 200000 ha under ESA agreement), considerable resources have been invested in monitoring the environmental impacts of the ESA schemes (Hooper, 1992).

Within the UK there are, therefore, two broad policy instruments that can be used to promote the deintensification of agricultural production. First, those associated with the CAP reforms, in particular the set-aside provisions of the Arable Area Payments scheme. Second, the designation of areas such as ESAs within which farmers are compensated for returning to traditional, less-intensive methods of production. These two policies both have the effect of reducing the intensity of agricultural production on the land directly designated, but it is important to realize that the policies have been created to satisfy different objectives. Set-aside is designed primarily to reduce agricultural surpluses, whereas ESAs and NSAs are designed to protect the environment.

The area of eligible land that can be placed under compulsory set-aside is currently between 15 and 18%, but may increase in future reviews of the CAP. Taking such large areas of land out of agricultural production ought to provide significant environmental benefits, but the present regulations have been designed primarily to reduce agricultural surpluses, and opportunities for environmental benefits are being lost as a result of two aspects of set-aside policy. First, the set-aside regulations do not prescribe management practices which foster environmental benefits. The annual mowing of rotational set-aside, for example, is designed to reduce problems caused by arable weeds, but if the mowing is carried out too early in the season the nests of breeding birds may be destroyed (Firbank *et al.*, 1993). Second, the selection of land parcels is generally undertaken in isolation by individual farmers; this results in the set-aside land being scattered thinly across the landscape, rather than in discrete blocks which would provide greater environmental benefits. Set-aside policy could be altered so that there is, for example, greater connectivity between the land parcels placed into set-aside, thus increasing their wildlife value. Furthermore, the management of such land would also need to change in order to maximize the potential environmental benefits. In particular, land would need to be taken out of intensive agricultural production on a semipermanent basis, rather than the largely temporary set-aside currently used. Some of these issues are now being addressed in the new MAFF Habitats Scheme (MAFF, 1994), in particular the pilot schemes for river margins land. Under this scheme farmers will enter into long-term agreements (20 years) to take land near river courses out of intensive arable agriculture, and manage it using low-intensity methods, in return for MAFF compensation payments. The pilot scheme is initially being run on a small-scale in six river systems in England.

In the case of schemes such as ESAs, the landscape which is to be managed for environmental benefits is designated by MAFF at the initial consultation stage. Farmers enter into long-term ESA agreements with their local MAFF officers, and use agricultural practices which are expected to benefit wildlife and nature conservation. These schemes have only been in place since 1987, and it is at present unclear whether the changes in farming practices being implemented will produce the desired environmental changes. Similarly, there is a need to predict the long-term economic impact of such policy changes, particularly with regard to farm income and labour requirements. Indeed, it will only then be possible to assess whether the most appropriate areas of land have been designated for inclusion in the schemes.

This chapter explores the issues surrounding the deintensification of agriculture, both to reduce agricultural surpluses and to protect the environment. This is in the context of two case studies. First, the ecological, economic and hydrological effects of set-aside policy in the catchment of the River Cam south of Cambridge are examined. Methods by which set-aside land could be targeted into areas where the environmental benefits would be maximal are considered, rather than simply using set-aside as a means of reducing surpluses. Specifically, the possible effects of the introduction of a form of targeted, semipermanent set-aside as part of a river corridor habitat management programme are considered. Second, the ecological and economic consequences of implementing ESA agricultural practices in the landscape of the Pennine Dales ESA in northern England are investigated, and the potential for the management changes imposed under typical ESA agreements to result in the desired environmental changes are studied. Both case studies utilize the integrated suite of models described in Chapters 2–4.

6.2 Deintensification of agriculture in the catchment of the River Cam

6.2.1 ENVIRONMENTAL EFFECTS OF ROTATIONAL SET-ASIDE

In the catchment of the River Cam, the landscape matrix is dominated by intensive arable agriculture, which occupies over 80% of the catchment area. Cereals, in particular winter wheat, are the dominant crop, but oilseed rape, root crops (potatoes, sugar beet) and protein crops (field beans) are also significant. The land is agriculturally very productive, with over 85% of the catchment being MLURI Land Capability Class 1 or Class 2.

Under the 1992 reform of the CAP, farmers must place some of their land under set-aside in order to qualify for Arable Area Payments (AAPs). This is the so-called 'compulsory' set-aside, and currently represents 15% of the farm land under a 6 year rotational agreement, or 18% of the land under a non-rotational agreement. It should be noted that the non-rotational set-aside may be in place for as little as 5 years under current regulations. Farmers may also volunteer to take additional agricultural land out of production.

6.2.2 TARGETED, SEMIPERMANENT SET-ASIDE IN LOWLAND RIVER CORRIDORS

Within the relatively uniform landscape of the catchment of the River Cam, river corridors have the potential to serve a number of important landscape functions. For example, seminatural riparian vegetation can provide a valuable resource for amenity and nature conservation in landscapes otherwise lacking extensive or continuous tracks of seminatural vegetation. The type of intensive arable agricultural production occurring in the catchment is associated with high levels of fertilizer usage (especially nitrogen) resulting in potential contamination of the ground- and surface-waters. It is now recognized that suitably vegetated 'buffer zones' in river corridors can protect in-stream water-quality, not only from fertilizer contamination, but also from excess sediment run-off. The habitat management of seminatural vegetation in river corridors is likely to entail either the removal of land from agricultural production, or else deintensification from the current arable agriculture. The potential economic effects of such landscape changes are therefore considerable.

Removing land along river corridors from intensive arable agriculture is effectively a form of targeted, semipermanent set-aside. Within the landscape as a whole, 15% of the arable land could be put into a river corridor scheme, a policy which would reduce agricultural surpluses in a similar way to conventional rotational set-aside. A major difference, of course, is that the amount of land taken out of intensive agriculture will vary considerably between individual farms, depending on their proximity to the river corridor areas. Furthermore, there are a number of advantages if the deintensified agricultural land is targeted along the river corridors, instead of being scattered thinly over the landscape. In particular, there are potential benefits, with appropriate long-term management, for amenity, nature conservation and hydrological protection.

(a) Amenity

River corridors can potentially form attractive seminatural landscape features in areas otherwise lacking such features. They are therefore of particular value near or within urban population centres. For example, in Greater London, river corridors have been developed to provide a valuable resource for recreation (Green, 1984). Intensively managed agricultural lands are often lacking in amenity value, with few footpaths, and only isolated areas of attractive seminatural vegetation within a landscape dominated by large arable fields (Cambridgeshire County Council, 1991). A more attractive landscape can be developed by habitat management of river corridors, linked to extant areas of seminatural vegetation, so that the river corridors contain a mixture of meadows, copses and marginal vegetation (Large and Petts, 1994). The resulting landscapes not only have greater amenity value in being more attractive, but they can also be more accessible, through the development of networks of paths along the river corridors.

(b) Nature conservation

Numerous studies have emphasized the importance of river corridors for wildlife, in particular riparian birds, mammals, aquatic macrophytes and invertebrates (Brooker and Welsh, 1982; Holmes, 1983; Rushton et al., 1994). In many intensively managed lowland landscapes, however, the nature conservation value of the rivers and streams has been severely reduced. At its most extreme, the original riparian vegetation has been destroyed and arable land borders the rivers directly. Such rivers have been characterized as acting simply as the gutters of the modern, intensively managed catchment. Unsympathetic river engineering schemes, often designed to alleviate flooding, have frequently involved straightening the river, stripping away bankside vegetation, and removing obstacles to the flow of water. These schemes can result in the creation of rivers with trapezoidal channels and uniform banks, with low habitat diversity and poor species richness across all taxa (Newbold et al., 1983). The fact that such schemes have often been associated with the installation of land drains and the conversion of river corridor grazing pastures to arable agriculture has exacerbated the loss to wildlife.

The problems associated with such river systems, and the move towards less intensive agricultural production, have led to a growing awareness of the need to restore rivers to improve their wildlife potential. The aim of such habitat management is not to try to recreate the original riparian vegetation: such attempts are likely to fail because the hydrology will have fundamentally changed, and the nutrient loads (especially nitrogen and phosphorus) and sedimentation levels will be much higher (Higler, 1993). Furthermore, the species composition of the original riparian vegetation will invariably be unknown. Nevertheless, appropriate river corridor habitat management can be used to create new habitats, of considerable nature conservation potential, albeit differing from the original riparian habitats (Newbold et al., 1983). Appropriate habitat management methodologies for river corridors are gradually being developed, and implemented, by local planning authorities (Cambridgeshire County Council, 1991).

(c) Hydrological protection

During the last 20 years the concentration of pollutants, especially nitrates, in lowland rivers has increased, mainly as a result of intensive agricultural practices. The nitrate concentrations in many lowland rivers in southern England now exceed the EU limits for drinking water (Burt and Haycock, 1992). Point-source pollution control methods are not effective in isolation to control this pollution. Diffuse-source pollution is occurring because human activity is altering the structure of the landscape, such that the quantity of pollutants has increased. Therefore, landscape or catchment-level solutions are required to reduce the pollutants in the long term.

Two methods have been suggested to alleviate diffuse-source pollution (Haycock et al., 1993). The first is to deintensify agricultural production, and in particular to reduce the application of nitrate fertilizers. The UK government has recognized the potential of such approaches, with the introduction of the Nitrate Sensitive Area (NSA) schemes (Burt and Haycock, 1992), and the implementation of EU Drinking Water directives. The second approach is to target hydrological protection methods to specific areas of the catchment, in particular the river corridors (Haycock et al., 1993). Numerous studies have shown the potential of riparian vegetation 'buffer strips' or 'buffer zones' to reduce the nitrate concentrations of river water (e.g. Peterjohn and Correll, 1984; Muscutt et al., 1993). It has been suggested that the use of such buffer strips provides the most ecologically sound, sustainable and economic approach to protecting the water quality of lowland rivers. Whereas riparian buffer zones may reduce diffuse-source pollution of river waters, they will be less effective in alleviating diffuse-source pollution of groundwaters outside the immediate locality of the buffer zone. Groundwaters in the landscape as a whole are already heavily polluted with nitrates in many catchments, and a long-term strategy of reducing agricultural nitrate fertilizer applications is required.

The effectiveness of a buffer zone in lowering local rootzone nitrate concentrations is dependent on the plant uptake of nitrate and the denitrification rate in the soil column. Since denitrification requires anaerobic conditions, its rate is increased by soil waterlogging. This means that floodplains and the areas adjacent to rivers often provide favourable conditions for denitrification.

6.2.3 DEFINING THE RIVER CORRIDORS

It is difficult to define the exact width of a river corridor because the riparian vegetation often grades gradually into adjacent habitats (Gurnell et al., 1994). The corridor width will vary depending on the form of the landscape and the nature of adjacent vegetation (Petts, 1990; Large and Petts, 1994). Nevertheless, attempts have been made to formalize the definition of a river corridor, particularly for the purposes of survey work. The National Rivers Authority (1992) suggests that the width of the corridor should depend on how much the nearby land is affected by the river, and *vice versa*, but in general it recommends that the corridor be defined as being the land within 50 m of each river bank. Where, as in this case study, the river corridor is also to function as a hydrological buffer zone, ranges of between 10 and 150 m have been suggested for the corridor width (Haycock *et al.*, 1993). It is clear, however, that for such buffer zones to be effective, a mixture of vegetation habitat patches is desirable.

In this case study, the river corridor, therefore, is not defined as a simple 'ribbon' of vegetation of uniform width blending poorly into the landscape, but rather as a mosaic of habitat patches, of different sizes, adjacent to the river (Large and Petts, 1994). This has the advantage of avoiding any arbitrary definitions of width, and the increased diversity of habitats in the corridor will increase the nature conservation and amenity value.

The area chosen for this case study was the 25 km × 25 km square of land due south of the city of Cambridge (NGR North TL259000, NGR South TL249000, NGR East TL550000, NGR West TL540000). Figure 6.1 shows the river network in this area: the rivers Cam, Rhee and

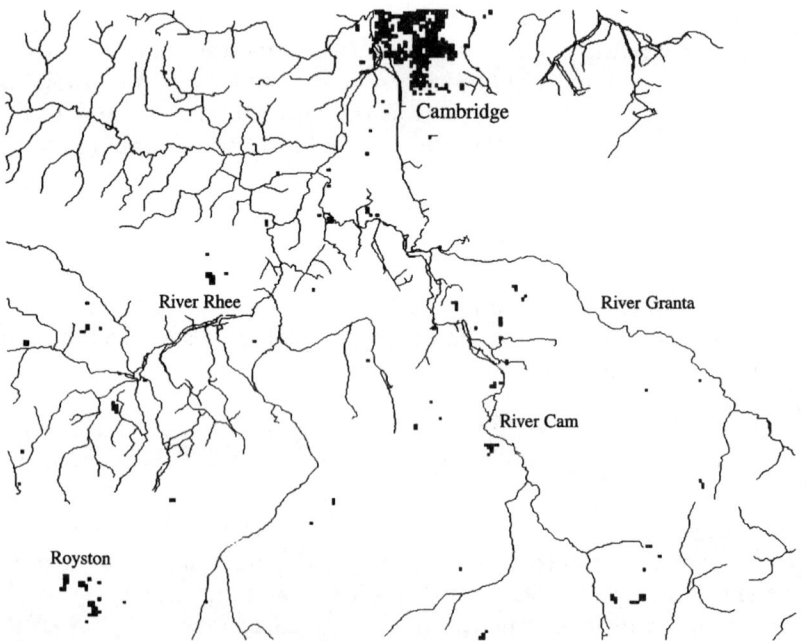

Figure 6.1 River network south of Cambridge.

Granta flow northwards, with the confluence of these rivers being c.5 km south of Cambridge. The area is currently dominated by arable agriculture.

A three-stage GIS procedure was used in the initial selection of the river corridors. Maps were manipulated within a raster-based Geographic Information System (GRASS) at a resolution of 25 m:

1. All blocks of non-arable or urban land within 2 km of the river were considered as potential corridor components. This was to define the outer limits of any habitat patches that could potentially be included in the corridor.
2. All land (including arable areas) within 100 m of the river was selected. This stage in the procedure was to define the 'inner core' of the river corridor, so that there would be continuity in the riparian vegetation along the river. This continuity would improve the value of the river corridor for amenity, for example by providing a complete network of footpaths. In addition, the continuity would improve the hydrological buffering capacity, and increase the value of the river corridor for wildlife.
3. Finally, the remaining areas in the 2 km buffered zone were clumped into discrete patches, and any clumps which did not overlap the 100 m buffered zone were deleted. The resulting map is shown in Figure 6.2. The river corridors shown in this map contain a mixture of habitats, across five major soil types.

6.2.4 ECOLOGICAL, ECONOMIC AND HYDROLOGICAL MANAGEMENT AND PREDICTIONS

(a) Ecology of the river corridors

The effects of the proposed river corridor scheme on the ecology of the terrestrial vegetation and riparian birds were investigated using the modelling approaches outlined in Chapter 4. The vegetation is a key component of most terrestrial ecosystems, effectively defining the biotic habitat for invertebrates, birds and mammals. Riparian birds generally have a high public profile, as well as being relatively sensitive to changes in the landscape.

Terrestrial vegetation

The aim in river corridor habitat management is not to create one single, homogeneous vegetation type along the corridor, but rather, to create a diversity of habitats. Therefore, the management of any selected areas that were currently deciduous woodland would be

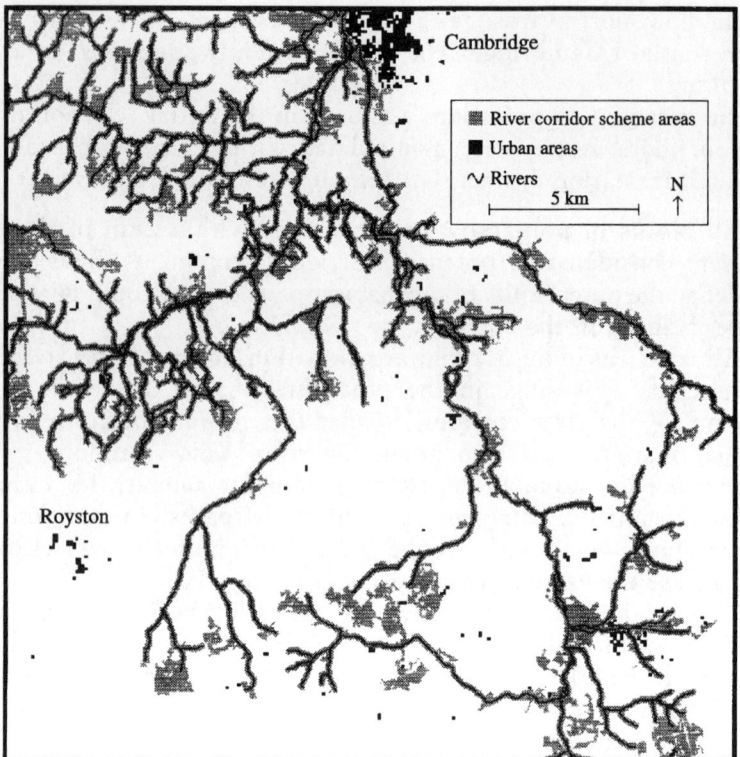

Figure 6.2 Location of the proposed river corridor habitat management scheme.

broadly unchanged, to retain habitats of potentially high wildlife value. Areas of arable land (within 100m of the river bank) and intensively managed pasture or amenity grass would need lower intensity management to enhance their wildlife value. It was assumed that these areas would be managed as low input grazing pastures, especially hay-meadows, with restrictions on stocking rates and inorganic fertilizer applications, or (on chalk rendzina soils) extensive year-round sheep grazing. The potential vegetation communities that would develop under such management were predicted using the vegetation model developed by Sanderson *et al.* (1995). Table 6.1 summarizes the results of the vegetation model, showing the predicted dominant plant communities of the National Vegetation Classification (Rodwell, 1992). The *Festuca ovina–Avenula pratensis* (CG2) grassland is predicted to be dominant on the rendzina soils; this is a very species-rich plant community, which has been in decline in recent years. Hay meadow communities (MG5) are likely to be dominant on the brown earths,

Table 6.1 Predicted dominant vegetation types within river corridor scheme, following a change to low-intensity pastoral agriculture, listed by soil type. Vegetation codes follow the National Vegetation Classification (Rodwell, 1992): CG2 *Festuca ovina–Avenula pratensis* grassland, CG3 *Bromus erectus* grassland, CG4 *Brachypodium pinnatum* grassland, MG5 *Cynosurus cristatus–Centaurea nigra* grassland, MG11 *Festuca rubra–Agrostis stolonifera–Potentilla anserina* grassland, MG13 *Agrostis stolonifera–Alopecurus geniculatus* grassland, M22 *Juncus subnodulosus–Cirsium palustre* fen-meadow

Soil Type	*Percentage of river corridor*	*Dominant vegetation communities*
Rendzina	10	CG2, CG3
Brown earth	7	MG5, MG11
Brown calcareous earth	33	MG5, CG4
Ground-water gley	11	MG11, MG13
Pelosol	40	MG11, M22

whereas on the wetter soils the *Agrostis stolonifera–Alopecurus geniculatus* (MG13) grassland, which was once a characteristic fenland grassland, is likely to develop. These grassland communities would be intermixed with small stands of currently extant woodland within the corridor area.

Riparian birds

The modelling approach developed by Rushton *et al.* (1994) was used to investigate the change in the habitat suitability for riparian birds after the introduction of a river corridor scheme. Figure 6.3 shows the change in predicted habitat suitability for two species of birds characteristic of high quality lowland river systems, the yellow wagtail (*Motacilla flava*) and the grasshopper warbler (*Locustella naevia*). The change in the habitat suitability for the yellow wagtail is variable: along the main river channels there is little predicted change, or even a slight decrease in the suitability. The habitat suitability, however, increases along many of the minor tributaries following the land use change. these are river corridors which were dominated by intensive arable agriculture before the land use change, and the creation of low-intensity management grasslands will provide additional habitats for nesting and feeding. In contrast, the habitat suitability for the less common grasshopper warbler is expected to increase across the whole of the river network following the introduction of a river corridor scheme.

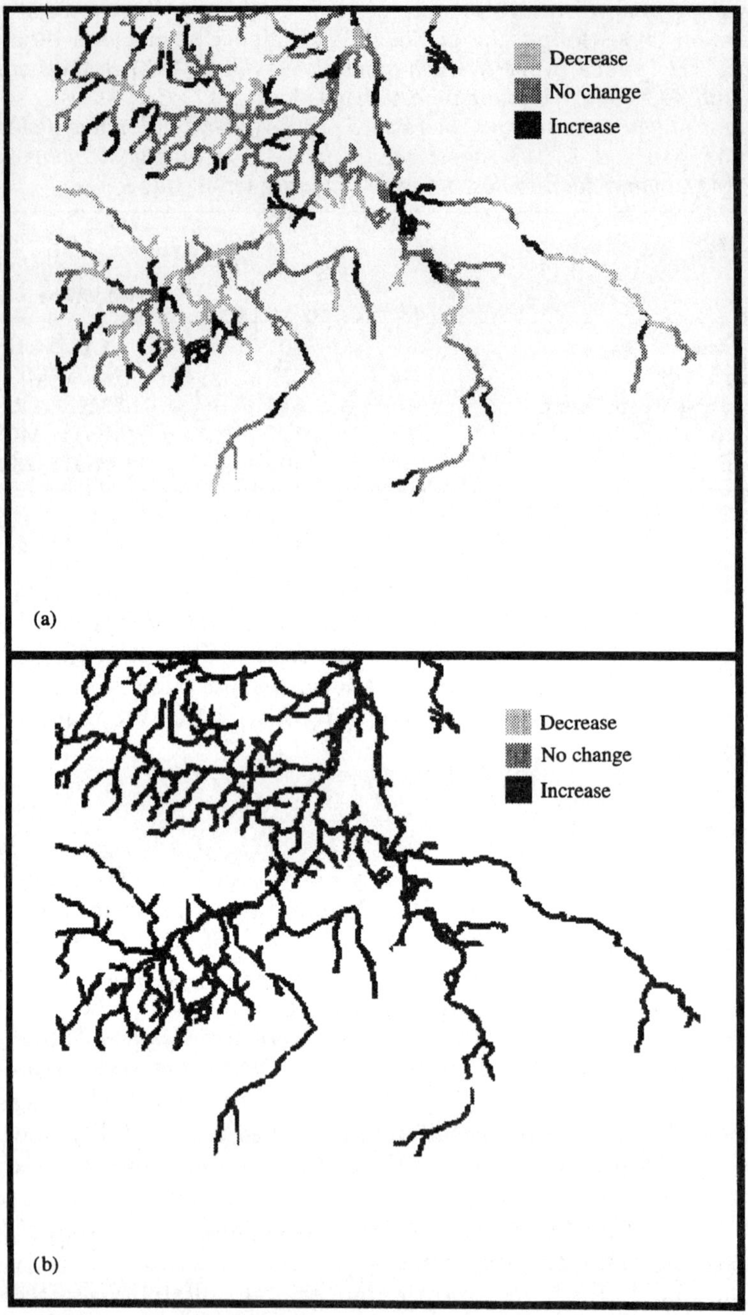

142 Deintensification of agriculture

Figure 6.3 Change in the predicted habitat suitability following the introduction of the river corridor scheme for (a) yellow wagtail. (b) grasshopper warbler.

Table 6.2 Economic impact of river corridor habitat management

	Base	River corridor	% change
Nitrate fertilizer costs (£ m)	27.9	25.0	10.4
Machinery costs (£ m)	19.9	19.2	3.5
Regular labour costs (£ m)	22.4	21.8	2.7
Total profit	90.6	88.4	2.4

Profit reduction within corridor scheme (£/ha): 242.

(b) Economic effects of river corridor habitat management

The economic impact of the proposed habitat management scheme was explored using the catchment-level linear programming model described in Chapter 2; the results are summarized in Table 6.2. The land use change is not, of course, evenly distributed across the landscape, but it is clear that there will be an impact on farm economics outside the immediate locality of the river corridor. The habitat management will result in an increase in the amount of low-intensity grassland, and as there is an EU limit on the number of livestock, there will be an excess of grazing pastures. Therefore, the management of existing grassland will be less intense, with lower inputs of nitrogen, even in grasslands outside the river corridor area. Within the river corridor area, there is a reduction of profit of £242/ha; in contrast, the pilot MAFF River Margins Scheme (MAFF, 1994) is adopting payments of £240/ha on permanent grass, and £360/ha on arable land. Any compensation, however, would have to recognize the increased costs to the individual farmer in the early years of entering the scheme, partly as a result of fixed costs.

(c) Effects on river water quality

The hydrological effects of the proposed river corridor habitat management scheme were investigated using a small-scale SHETRAN model of a representative block of land alongside the river. An area 500m by 500m was modelled with a grid resolution of 50m. The physical characteristics of the model were based on the area along the Rivers Cam and Rhee; the topography of the model is gently sloping down to the river and the subsurface is clay loam soil overlying some 15m of chalk. Boundary conditions describing the groundwater flow and nitrate concentrations passing into the block of land were estimated from the physical attributes of the upstream catchment. Prior to the river corridor habitat management scheme, the block of land was assumed to be under winter wheat and to have an annual nitrate

144 Deintensification of agriculture

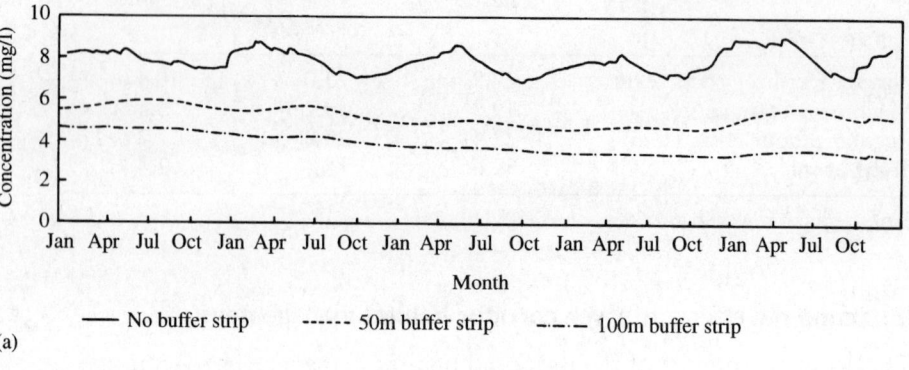
(a)

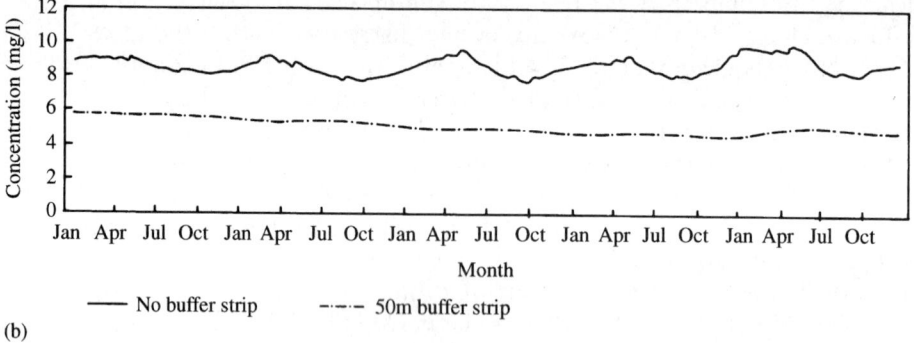
(b)

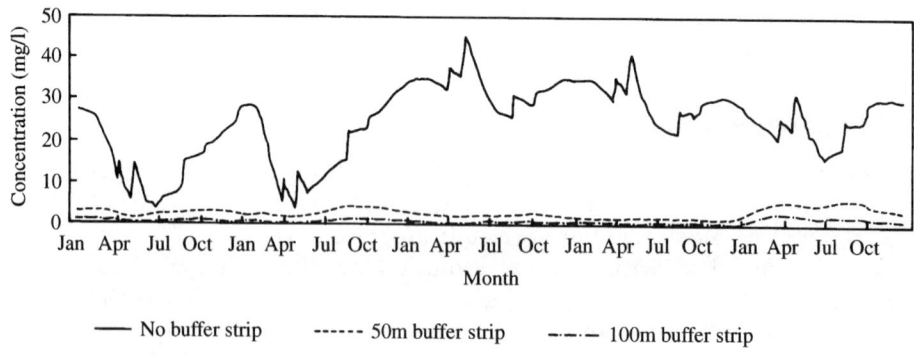
(c)

Figure 6.4 Effects of the river corridor scheme on water quality. (a) Groundwater nitrate concentration 0–50 m from river; (b) groundwater nitrate concentration 50–100 m from river; (c) concentration of nitrate in leachate adjacent to river.

fertilizer application rate of 200kg/ha. Predictions were made of the movements of nitrate under this land use using a 5-year time-series of meteorological data from 1989 to 1993. The concentration of nitrate discharged into the river is governed by the nitrate concentration in the surface and groundwater along the river bank. For this example, the surface water flows were found to be negligible, so the nitrate concentration in the river is influenced purely by the adjacent groundwater concentration. To investigate how the nitrate concentration might be lowered by the use of a buffer strip, a strip of land along the river bank was removed from arable cultivation and converted to low nitrate input grassland. Two different widths of buffer strip (50m and 100m) were tested, at both high and low denitrification rates.

Results from the first set of hydrological simulations, which assumed a low denitrification rate, are shown in Figure 6.4. Figure 6.4(a) compares groundwater nitrate concentrations immediately adjacent to the river for three scenarios: no buffer strip, 50m buffer strip and 100 m buffer strip. It can be seen that a 50m buffer strip lowers the groundwater nitrate concentration by around 30%, whereas a 100m buffer strip lowers the concentration by around 50%. In the next 50m away from the river, the groundwater concentration is also reduced by around 30% when a 100m buffer strip is used (Figure 6.4b). Figure 6.4(c) shows the differences in the nitrate concentration of the leachate in the strip of elements adjacent to the river under the different scenarios. The peaks in the concentration of leachate correspond to the release of nitrate from the fertilizer applications in the spring. As expected there is a large reduction in the concentration of the leached nitrate without any fertilizer application on the buffer strip.

Table 6.3 summarizes the differences that were observed when a high denitrification rate was assumed. With the area under winter wheat, the peak nitrate concentration of the leachate is approximately 70% of that

Table 6.3 Average nitrate concentrations under conditions of high and low denitrification.

	High denitrification ($2.78e^{-08} s^{-1}$)	Low denitrification ($2.78e^{-09} s^{-1}$)
Average groundwater concentration: no buffer strip	1.2 mg/l	8 mg/l
Average groundwater concentration: 50m buffer strip	0.3 mg/l	5 mg/l
Average groundwater concentration: 100m buffer strip	0.1 mg/l	3 mg/l
Peak concentration of leachate	35 mg/l	50 mg/l

estimated using a low denitrification rate. The corresponding ground water nitrate concentration is reduced by an even higher margin from around 8 mg/l to just over 1 mg/l. When a 50 m buffer strip is imposed, the ground water concentration adjacent to the river is reduced to only 0.3 mg/l. Thus, with a high denitrification rate the effectiveness of the buffer strip is greater, since it reduces the ground water concentration by a higher percentage.

In summary, the predictions from the hydrological modelling show that the proposed river corridor habitat management scheme would be extremely effective in lowering the concentration of nitrates discharged into the rivers, hence improving the general level of the river water quality. A buffer strip 50 m wide is sufficient to cause a significant reduction in the concentration of nitrates, but a 100 m strip is even more successful. The effectiveness of the buffer strip will be greatest where local conditions encourage a high denitrification rate. Although this scheme should be successful in improving river water quality, it does not, of course, solve any problems associated with the general level of ground water quality in the catchment. This issue is perhaps equally as important as the surface water issue in the Cam region, since ground water is the main source for the public water supply. Improvements to the general level of ground water quality can only be achieved by reducing the agricultural applications of nitrate fertilizer.

6.3 Deintensification of agriculture in the Pennine Dales Environmentally Sensitive Area

6.3.1 INTRODUCTION

The Pennine Dales ESA was designated in 1987 by MAFF to protect the unique landscape which has developed in the area over centuries of low-intensity agriculture (MAFF, 1992; Saunders, 1994). One of the main objectives of the Pennine Dales ESA is to maintain and enhance the floral diversity of the grasslands (especially species-rich hay-meadows) which are the product of traditional management: this management involves the use of enclosed lowland for permanent pasture and hay production, with low inputs of inorganic fertilizer. The scheme was extended in 1992 to include additional dales, increasing the area of the Pennine Dales ESA by approximately 30%. The ESA scheme is administered by MAFF regional and divisional staff with first-hand experience of ecological and agricultural issues. ESA agreements for individual farms are usually made between the farmer and his ESA Project Officer, within the framework of the published ESA guidelines (MAFF, 1992), with the result that there is often minor variation in the way in which ESA agreements are implemented between different farms. The aim of this case study is to investigate the economic and ecological effects at the farm level of entering into a Pennine Dales ESA-type agreement. In addition, the ecological effects of implementing ESA management prescriptions on species distributions in landscapes are explored. The ESA areas have already been designated, and modelling systems can be used to examine whether such highly specific policies are at least likely to achieve their environmental goals, within acceptable economic limits.

6.3.2 MANAGEMENT OF GRASSLANDS UNDER THE PENNINE DALES ESA

The management prescriptions for the Pennine Dales ESA are concerned with maintaining the structure of the landscape associated with pastoral agriculture, specifically in the form of a traditional, relatively low-input agricultural system. Management is aimed primarily at maintaining herb-rich grassland through restricting the use of inorganic or natural fertilizers and through management of the way in which herbage is utilized. The specific ecological objectives are to minimize reductions in sward diversity arising from competition from highly productive species such as *Lolium perenne* and to ensure that the herbage is utilized only after herb species have flowered and set-seed. In return for following these prescriptions, farmers receive a yearly cash payment of £125/ha for areas managed as meadows and £95/ha for other types of grassland (Saunders, 1994). A second tier to the scheme was included in the 1992 revision of the ESA, where farmers received higher payments (£210/ha) in return for stricter controls on fertilizers and cutting dates (Saunders, 1994).

The management prescriptions can be listed as follows.

1. *Sward disturbance and utilization regime:*
   - Abstention from ploughing, reseeding or cultivation of grassland;
   - Stock to be excluded from meadows for at least 7 weeks before the first hay cut (or 1 June at latest);
   - Grass must not be cut before 8 July;
   - All meadows must be cut first in August once every 5 years;
   - New drainage not to be installed;
   - Pastures must not be overgrazed, poached or undergrazed.
2. *Sward fertilizer regime:*
   - Current inorganic fertilizer must not be exceeded and a maximum of 25 kg/ha nitrogen, 12.5 kg/ha each of potassium and phosphate in artificial fertilizer can only be applied;

- Fertilizer must be applied all at once;
- Slurry or poultry manure must not be used;
- Current use of farmyard manure must not be exceeded and less than 12.5 t/ha applied;
- Manure applied in one batch;
- Slag or lime to increase soil pH must not be applied;
- Fungicides and insecticides must not be used.

Three of the dales within the Pennine Dales ESA fall within the catchment of the River Tyne; the South Tyne with Nent Valley, the West Allen Dale and the East Allen Dale. These dales were added to the ESA in the 1992 revision of the scheme, and the economic and ecological effects of ESA management were investigated using the model system described in Chapters 2 and 4.

6.3.3 ECONOMIC EFFECTS OF ESA PRESCRIPTIONS

The Catchment-Level LP model within the DSS (Chapter 2) was used to investigate the economic effects of reduced intensity of management as required under ESA schemes. Under the Tier 1 proposals of the ESA, the model predicted that the introduction of the ESA would result in an additional farm income of £113/ha. The area under silage production declines sharply within the ESA (Table 6.4), whereas the area under hay meadow management increases. It is possible that farmers outside the ESA might sell excess forage to those farms under ESA management (or even increase production) with the development of a local forage market. Overall, therefore, it appears that farmers will be slightly overcompensated by the introduction of the ESA scheme. Similar results were obtained by Oglethorpe *et al.* (1995) when the possible economic effects of ESA management were examined at the farm-level.

Table 6.4 Predicted effects of introduction of ESA scheme in the Pennine Dales (South Tyne) on land useage

Area (thousands of ha)	Pre-ESA	Post-ESA
Hay meadow	2.1	8.7
Silage production	19.6	12.4
Permanent pasture	21.2	21.2

Additional farm income (£/ha) post-ESA: 113.4.

6.3.4 ECOLOGICAL EFFECTS OF ESA PRESCRIPTIONS

The major objective of adopting ESA management prescriptions is to stop reseeding, reduce fertilizer applications and maintain hay-meadow management on established pasture. These changes were incorporated in the DSS in the form of transitions from intensively managed pastures and good semi-improved pasture to poor semi-improved land cover. As with the economic models, ESA prescriptions against over- and undergrazing and poaching were not incorporated explicitly into the analysis.

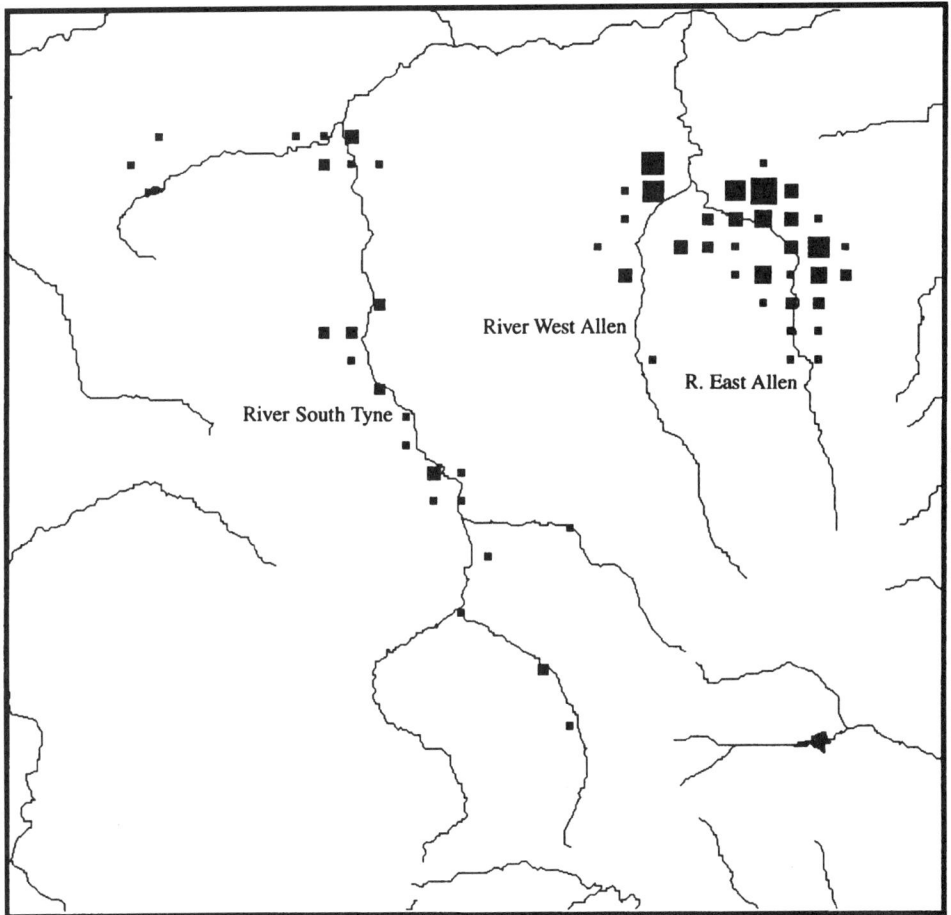

Figure 6.5 Increase in the habitat suitability for hay-meadow community MG3 following the introduction of Pennine Dales ESA management. Area shown is the catchment of the South Tyne and West and East Allen rivers.

(a) Plants

An ecological feature of particular interest in the northern part of the Pennine Dales ESA is the presence of herb-rich hay-meadow communities such as the *Anthoxanthum odoratum–Geranium sylvaticum* grassland community (MG3) of the National Vegetation Classification (Rodwell, 1992). The predicted change in distribution of this community after the adoption of ESA prescriptions was investigated in order to assess the extent to which the ESA prescriptions would be successful if applied in the three dales in the study area.

The predicted change in the distribution of the *Anthoxanthum odoratum-Geranium sylvaticum* community is shown in Figure 6.5. The community is predicted to become more common along the length of the South Tyne, although the increase is slight. A more marked increase is predicted in the Allen Dales, especially in East Allen Dale, and at the confluence of the West and East River Allens. These relatively wide, flat-bottom dales contain large areas of in-bye land,

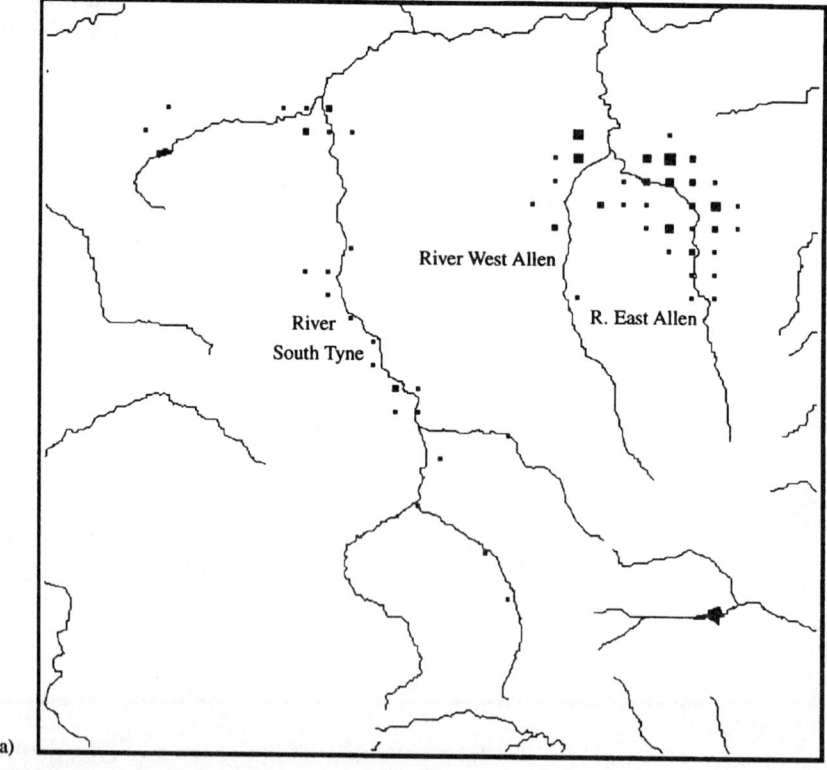

Figure 6.6 Increase in the habitat suitability for birds following the introduction of Pennine Dales ESA management: (a) snipe, (b) lapwing.

which is likely to respond to a change in management towards more traditional, hay-making techniques.

(b) Birds

The birds which will be most affected by the adoption of ESA prescriptions are those which breed in agricultural grasslands. Few species are able to nest successfully on intensively managed pasture. Considered here are two species which are in decline as breeding birds in the British Isles as a result of the intensification of agriculture in the lowlands; the lapwing (*Vanellus vanellus*) and the snipe (*Gallinago gallinago*). Figure 6.6 shows the predicted change in the distribution of these species in the three dales as a result of adopting ESA management. Both species show an increase along the valley bottoms, with the biggest changes in the West and East Allen Dales.

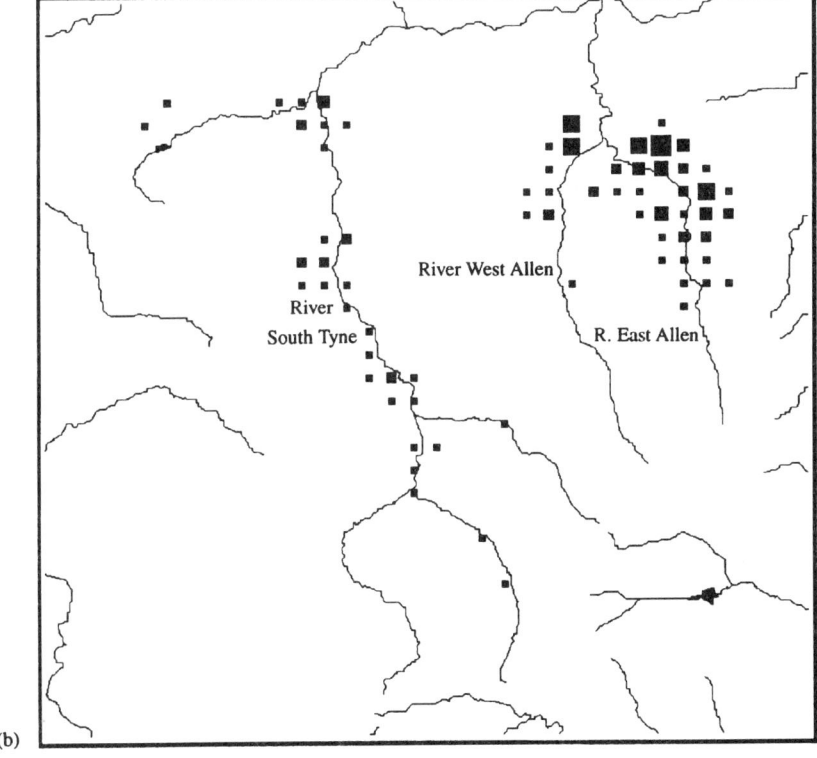

Figure 6.6 continued

6.3.5 IMPLICATIONS OF ESA POLICY

This landscape-scale analysis suggests that the adoption of Pennine Dales ESA management policies should provide some of the desired environmental benefits. Similar studies at the smaller resolution of the farm-scale on the economic and ecological effects of ESA agreements also predict beneficial environmental changes (Oglethorpe et al., 1995). It should be noted, however, that no time-scale has been placed on these predictions, and some of the ecological changes may take many years. The Pennine Dales ESA is unusual in that unlike most ESAs it is fragmented, currently comprising 20 discrete blocks in different dales, and the linear shape of these blocks results in many farms having parts of their land outside the ESA (Saunders, 1994). When the ESA was created there were concerns that this might lead to a 'halo effect' in which farmers increased the intensity of production on those parts of the farm outside the ESA, to compensate for reduced productivity within the ESA. In practice this does not appear to have been a problem, although there has been a slight increase in fertilizer usage outside the ESA (Saunders, 1994).

6.4 Conclusions: land-use policy and agricultural deintensification

This chapter has examined some of the issues associated with designating land for agricultural deintensification, and the management necessary to provide the desired environmental benefits. The river corridor scenario proposed for the catchment of the River Cam involved targeting the set-aside land into specific areas, on a semipermanent basis. The total area of arable land included was the same proportion (15%) as that which can be set-aside under current (per farm) regulations, yet the likely effects on the landscape would be dramatically different. New habitats would be created, and there would be considerable benefits for wildlife, amenity and water quality. One effect of this approach is that farms near the affected rivers might have to set-aside a major part of their arable land into the scheme, whereas farms away from the rivers might be virtually unaffected.

The implications of the policy changes necessary to achieve such landscape change are immense. It is unlikely that it would ever be politically or socially acceptable to compel farmers or landowners within the river corridor to enter parts of their land into the scheme. Instead, there could be a sliding scale of compensation payments under the AAP system, with farms near the rivers being eligible for significantly higher payments. The recent changes in the regulations allowing farmers to 'trade' their set-aside land between different farms might make it easier to introduce forms of targeted semipermanent set-aside. The effects of such changes are complex and difficult to predict. For example, a high level of payments might result in farmers being able to invest in the equipment or labour necessary to increase production in that part of their land which was not covered by the river corridor scheme. Initial concerns, however, that the introduction of the current set-aside scheme might lead to farmers managing their remaining land more intensively have not so far been realized (Ansell and Tranter, 1992). In fact, the results from the linear programme presented here suggest that there might be a reduction in the intensity of management of grasslands outside the river corridors, while the livestock numbers remain fixed. Some of the river corridor land could be managed non-agriculturally, for example by applying various types of mowing regimes, but the resulting vegetation will generally be less species-rich in comparison with that under grazing management. The creation of a diversity of habitats within the river corridor should be an aim of such schemes, hence the importance, for example, of retaining intact existing areas of woodland within the river corridor.

In contrast, in the case of the ESAs, the land has already been designated for inclusion within the schemes. The number of ESAs in the UK was substantially increased in 1992 to almost 2 million hectares (Whitby and Lowe, 1994), although it is unlikely that there will be any further extensions or additional new ESAs in the immediate future. The three dales considered in this chapter were added to the Pennine Dales ESA as part of the 1992 revision. Both the economic and ecological models suggest that although the ESA is unlikely to result in any massive land use changes, those changes which do occur will be environmentally beneficial. The area is already subject to relatively low intensity agricultural production, and the ESA scheme will serve to ensure that such traditional methods of farming continue to survive.

In both the river corridor and ESA examples, it is clear that policy incentives to deintensify farming should provide significant environmental benefits in the long term. It is, however, more difficult to quantify the socioeconomic implications of such policy changes. Although the ESA schemes are generally regarded as a 'success', reflected by their high uptake amongst farmers (Whitby and Lowe, 1994), the radical changes to set-aside policy suggested for the river corridor scheme would be highly controversial. Any revision of the current set-aside policy would affect a much larger area and more farmers than in the ESA schemes. However, policy makers need to address the issue of how to modify the set-aside provisions to provide environmental benefits without alienating the farming community.

References

Ansell, D.J. and Tranter, R.B. (1992) *Set-aside: in Theory and Practice*. Centre for Agricultural Strategy, University of Reading.

Brooker, M.P. and Welsh, W.A. (eds) (1982) *Conservation of Wildlife in River Corridors*, UWIST/Welsh Water Authority.

Burt, T.P. and Haycock, N.E. (1992) Catchment planning and the nitrate issue – a UK perspective. *Progress in Physical Geography*, **16**, 379–404.

Cambridgeshire County Council (1991) *Cambridgeshire Landscape Guidelines: a Manual for Management and Change in the Rural Landscape*. CCC/Countryside Commission, Granta Editions.

Firbank, L.G., Arnold, H.R., Eversham, B.C. *et al.* (1993) *Managing Set-aside Land for Wildlife*, HMSO, London.

Green, B. (1984) *The Brent River Park: Report of Landscape Proposals*. London Borough of Ealing.

Gurnell, A., Angold, P. and Gregory, K.J. (1994) Classification of river corridors: issues to be addressed in developing an operation methodology. Aquatic conservation. *Marine and Freshwater Ecosystems*, **4**, 219–31.

Haycock, N.E., Pinay, G. and Walker, C. (1993) Nitrogen-retention in river corridors – a European perspective. *Ambio*, **22**, 340–6.

Higler, L.W.G. (1993) The riparian community of north-west European lowland streams. *Freshwater Biology*, **29**, 229–41.

Holmes, N.T.H. (1983) *Typing British rivers according to their flora. Focus on Nature Conservation*, number 4, Nature Conservancy Council, Peterborough.

Hooper, A.J. (1992) *Field monitoring of environmental change in the Environmentally Sensitive Areas*, in *Land Use Change: The Causes and Consequences* (ed. M.C. Whitby), HMSO, London.

Large, A.R.G. and Petts, G.E. (1994) Rehabilitation of river margins. In *The Rivers Handbook – Hydrological and Ecological Principles* (eds P. Callow and G.E. Petts), Vol. 2, Blackwell, Oxford, pp. 401–18.

MAFF (1992) *The Pennine Dales Environmentally Sensitive Area: Guidelines for Farmers*. Ministry of Agriculture, Fisheries and Food, PD/ESA/2, London.

MAFF (1994) *The Habitat Scheme: Water Fringe Areas*. Ministry of Agriculture, Fisheries and Food, HS/WF/2, London.

Muscutt, A.D., Harris, G.L., Bailey, S.W. and Davies, D.B. (1993) Buffer zones to improve water quality: a review of their potential use in UK agriculture. *Agriculture, Ecosystems and Environment*, **45**, 59–77.

National Rivers Authority (1992) *River Corridor Surveys – Methods and Procedures*. Conservation Technical Handbook No. 1.

Newbold, C., Purseglove, J. and Holmes, N. (1983) *Nature Conservation and River Engineering*. Nature Conservancy Council, Peterborough.

Oglethorpe, D.R., Sanderson, R.A. and O'Callaghan, J.R. (1995) The economic and ecological impact at the farm level of adopting Pennine Dales Environmentally Sensitive Area (ESA) grassland management prescriptions. *Journal of Environmental Planning and Management*, **38**, 125–36.

Osborne, L.L., Bayley, P.B., Higler, L.W.G. *et al.* (1993) Restoration of lowland streams – a review. *Freshwater Biology*, **29**, 187–94.

Peterjohn, W.T. and Correll, D.L. (1984) Nutrient dynamics in agricultural watersheds: observations on the role of a riparian forest. *Ecology*, **65**, 1466–75.

Petts, G.E. (1990) The role of ecotones in aquatic landscape management, in *The Ecology and Management of Aquatic–Terrestrial Ecotones* (eds R.J. Naiman and H. Décamps), Parthenon Publishing, Carnforth, pp. 227–62.

Rodwell, J.S. (1992) *British Plant Communities*: Vol. 3, *Grasslands and Montane Communities*. Cambridge University Press, Cambridge.

Rushton, S.P., Hill, D. and Carter, S.P. (1994) The abundance of river corridor birds in relation to their habitats – a modelling approach. *Journal of Applied Ecology*, **31**, 313–28.

Sanderson, R.A., Rushton, S.P., Pickering, A.T. and Byrne, J.P. (1995) A preliminary method of predicting plant species distributions using the British National Vegetation Classification. *Journal of Environmental Management*, **43**, 265–88.

Saunders, C. (1994) Single-tier system with many farms partly outside the ESA: the case of the Pennine Dales, in *Incentives for Countryside Management: The Case of Environmentally Sensitive Areas* (ed. M.C. Whitby), CAB International, Wallingford.

Whitby, M. and Lowe, P. (1994) The political and economic roots of environmental policy in agriculture, in *Incentives for Countryside Management: The Case of Environmentally Sensitive Areas* (ed. M.C. Whitby), CAB International, Wallingford.

7 Forestry developments in the uplands and lowlands

7.1 Introduction

Since the establishment of the Forestry Commission in 1919, the percentage of the UK land area covered with trees has increased from 5% to 10%. This, however, represents a very small area in comparison with that which existed several hundred years ago and is considerably lower than most European countries. The native woodlands of the UK have gradually disappeared over the years as more and more land has been taken over for agricultural production. Now, however, the existence of agricultual surpluses means that pressure on the land is being reduced and the further expansion of forestry in the UK is set to continue. Additional factors motivating an increase in forestry include pressure from environmentalists to reintroduce native species to their natural areas, and the requirement to increase the national timber resource both for use as a raw material and also increasingly for energy production.

The vast majority of the forestry developments since 1919 have occurred in upland regions of the UK, primarily Scotland, Wales and northern England. Most of the trees planted in these regions have been fast-growing coniferous species, because the schemes have been geared to timber production and the generation of rural employment. As such, the developments have done little to restore native species and are considered by many to have had a detrimental effect on the quality of the landscape. However, in recent years a shift in forestry policy has occurred, towards multiple objective forestry development (Countryside Commission, 1993). The idea behind this is to encourage developments which have benefits in terms of the creation of wildlife habitats, improvements to the quality of the landscape and the provision of recreational areas for the public, in addition to timber production and employment. Under a joint initiative between the Countryside Commission and the Forestry Commission, a scheme was introduced in 1987 to create large areas of multipurpose Community Forests in and around suburban areas of the UK. Several of these developments are now under way.

Within this case study, two contrasting scenarios of afforestation are investigated using the NELUP DSS. The first considers a case of afforestation in an upland region. This scenario investigates the further expansion of the type of upland afforestation that has been common over the last 70 years, where the emphasis has been on large-scale timber production. The second scenario considers a case of afforestation in a lowland region where the aim is to develop a community forest, and thus aesthetic considerations are given as much weight as economic factors.

7.1.1 USE OF THE DSS IN FORESTRY CASE STUDIES

The data held within the DSS provide background information relating to the suitability of an area for different types of forestry. These data include:

- Climate variables – rainfall, temperature and wind speed;
- Topography – elevations and gradients;
- Soils data;
- Agricultural land capability.

The first three of these databases can be used to assess both the most appropriate locations for planting trees and also the most appropriate species of tree in the different locations. They can also be used to give an indication of how much land preparation and management would be necessary, e.g. whether trees would need protection because of the climate, how easy it would be to use machinery on the land, or whether improvements to the soil structure would be necessary. The agricultural land capability describes the quality of the land for agricultural production. It is derived from maps of soils, climate and topography using the methodology developed by the Macaulay Land Use Research Institute. The map of land capability can be used to provide an economic indicator of the cost of removing an area from agricultural production.

The base-line model simulations provide a description of the existing status of the region in terms of its hydrology, ecology and economics. Predictions from the base-line simulations may indicate sensitivity of the area to particular problems such as water shortages. Alternatively they may highlight important features such as the presence of rare plant or animal species. The proposed change in land use is applied to the models, and the results of the new simulations are evaluated by comparison with the baseline conditions. If the effects of the proposed change are undesirable, then modifications to the proposals may also be tested.

7.2 Upland forestry: the South Tyne

The western half of the Tyne Basin is an upland region, with rolling hills and elevations rising to around 550 m in the north of the catchment and 750 m in the south. Traditionally, the land has been used for hill-farming, with sheep the predominant source of income, although some of the land in the north is owned by the Ministry of Defence and is additionally used for military training, whereas estate-owned land in the south of the catchment provides grouse shooting. In the late 1940s and 1950s, a very large area of the North Tyne was planted for commercial forestry activities, creating the 250 km^2 Kielder Forest. Since then, many other areas of the North Pennines and the Border hills of southern Scotland have also been afforested. The hills in the South Tyne are very similar to the North Tyne in terms of their climate and the nature of the soils, and as such might be viewed as a possible area for further extending the forestry activities. Therefore, the aim of this first scenario is to investigate how an afforestation scheme in the south of the Tyne catchment would affect the characteristics of the region. The area investigated includes the subcatchment regions of the South Tyne at Alston, the South Tyne at Featherstone and the West Allen.

7.2.1 SOUTH TYNE BASE-LINE DATA

The region has relatively high rainfall, with an annual average of around 1500 mm, and there is little seasonal variation in the rainfall volume. Winter temperatures frequently fall below 0 °C, so there can be heavy snow at times. Although there is considerable variation in elevation, the topography is relatively smooth making large areas of the hill tops fairly exposed sites. The elevations range from 140 m to 750 m, with approximately 65% of the area below 450 m.

The Soil Association map shows that three soil types are dominant. On the highest areas, there is the Winter Hill association, which is a thick blanket peat. It has a high saturated water content and a tendency to remain waterlogged for much of the year. The Wilcocks association occurs on hill-slopes fringing the areas of Winter Hill. Wilcocks has a peaty surface layer overlying a clay loam, and again has a tendency to waterlogging. The third dominant soil type is the Brickfield association, which occurs on the lower areas. It is a clay loam which has a tendency to seasonal waterlogging.

The combination of high rainfall, hilly terrain and low permeability soils has resulted in a dense network of streams.

In terms of the agricultural land capability, all of the region is of poor quality, ranging from Class 3.2 to Class 6.3 – the lowest class defined. The areas of Class 6.3 correspond to the areas of the Winter Hill association, on the hill tops.

The inference of these baseline conditions is that an afforestation scheme in this region must be based around tree species that are hardy and that can withstand both high rainfall and poor quality soils. For coniferous trees this limits the choice largely to species such as sitka spruce, lodgepole pine, Norway spruce and Japanese larch. Any possibilities for including deciduous woodland would be limited to the most hardy species such as alder, birch and rowan. In terms of land preparation, the poor quality of the soils and their tendency to waterlogging means that intensive drainage would be necessary. This would probably involve the creation of a ridge–furrow type scheme, by deep ploughing. The trees would be planted on the drier elevated turf ridges between the ditches.

7.2.2 SOUTH TYNE BASE-LINE MODELS

(a) Hydrology

The base-line hydrology simulations show that the South Tyne is a region with a very high runoff ratio. This was an average of 0.76 for the period simulated, 1985–89. The runoff is dominated by surface flows rather than subsurface flows, with a ratio of around 3:1. This is a result of the combination of high rainfall on soils that have a tendency to waterlogging, and a very limited ground water system. As reflected by the high runoff ratio, losses to evapotranspiration are low, at around 400 mm per year. Since the runoff is dominated by surface flows, the response of the region to rainfall events is very fast, and the peak flows are high. For the South Tyne at Featherstone, the highest flows, which were exceeded only 0.1% of the time during the 5-year simulation, were greater than $343 \, m^3/s$. The low flows, exceeded 99.9% of the time, were greater than $2.2 \, m^3/s$. The predictions also show that the average level of the water table remains less than 0.5 m below the ground surface over around 95% of the area. This is an important issue with regard to potential afforestation as it indicates the need for deep ploughing and the implementation of drainage schemes prior to planting.

(b) Ecology

From the Landsat remotely sensed data, the South Tyne is seen to consist of a mixture of meadow/verge/seminatural areas in the valley bottoms, with moorland grass on higher areas, and a mixture of open and dense shrub moor on the hill tops. Predictions from the vegetation matrix model suggest that the shrub moor consists of considerable areas of *Calluna vulgaris*, or heather. The predicted probability of occurrence of *Calluna vulgaris* is shown in Figure 7.1(a). There is a high probability of occurrence of red grouse (Figure 7.1b), and this can be seen to be related to the presence of the heather moorland. Also of importance is the predicted occurrence of golden plover on the higher areas (Figure 7.1c). In terms of habitat, the predicted areas of bog shown in Figure 7.1(d) are also environmentally important.

(c) Economics

As shown by the land capability map, the South Tyne region is a marginal agricultural area, with some permanent pasture in the valley bottoms, and rough grazing on the hills. The presence of red grouse and heather moorland suggests that shooting may be a profitable activity. Possibilities for economic development are limited largely by the poor quality of the land.

7.2.3 IMPACTS OF SOUTH TYNE AFFORESTATION

As an initial assessment of the impacts on the South Tyne region, a blanket forest of coniferous trees was imposed on the area.

(a) Hydrology

The impacts of the blanket forestry on hydrology were investigated by considering the scheme at several stages through its life cycle. The most significant initial impact of an afforestation scheme in this area will be the effect of the initial land management and drainage. This can be investigated using an effective parameterization developed for representing the effects of open ditch drainage in SHETRAN (Dunn and Mackay, 1996). It was assumed that the drainage scheme would involve the creation of ditches 0.6 m deep, running down the line of the hillslope at intervals of 10 m. When the trees have grown to maturity, the hydrology of the area will be further modified by the change to the evapotranspiration regime that results from a change in land cover. At this stage it can be assumed that the original ditches will be partially filled in and that some of them will not have been maintained and no longer exist on the ground. To represent this, the depth of the ditches is reduced to 0.3 m deep and the density is reduced to one ditch every 20 m. The tree canopy is assumed to be 10 m high. Further changes to the hydrology will occur once felling of the forest begins to take place.

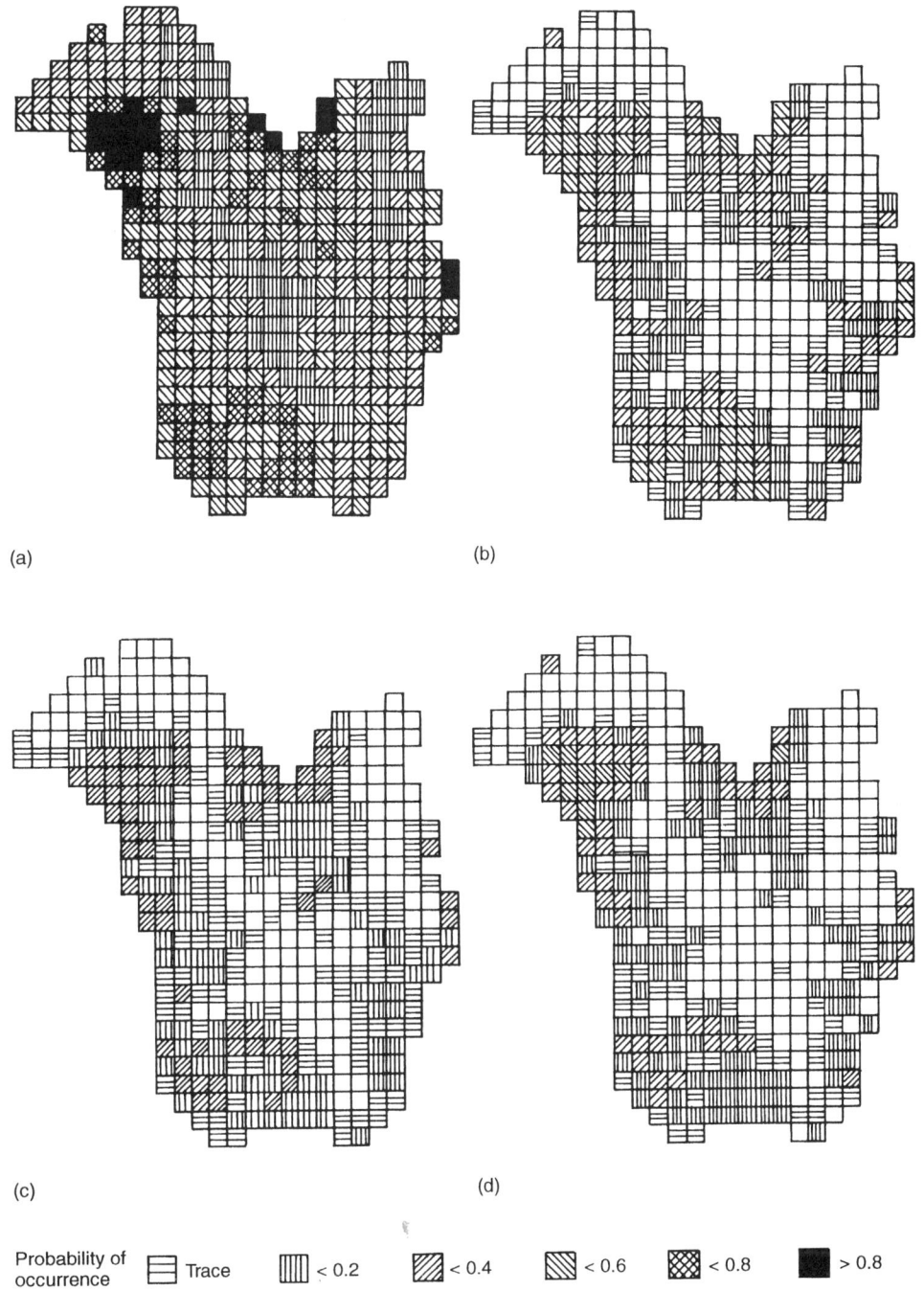

Figure 7.1 South Tyne: distribution of important ecological species. (a) heather; (b) red grouse; (c) golden plover; (d) blanket mire.

For each of these three stages a different SHETRAN simulation was performed.

Results from the simulations are compared with the base-line conditions to evaluate the impact of the change in land cover. Changes to the mass for the three simulations are summarized in Table 7.1, flow duration curve statistics can be compared in Table 7.2, and two sections of the hydrographs are illustrated in Figure 7.2(a, b).

In the first simulation, the main effect of the intensive drainage is to change the nature of the runoff regime, from surface water dominated to subsurface dominated. This is because the ditches create a large area of bisection with the saturated subsurface, thus increasing the exchange flow from the subsurface into the channel network. A corresponding lowering of the water table occurs. However, the time-delay caused by the increase in subsurface flow is minimal, because the subsurface travel distance to the ditches is small. The short distance to the ditches also has a secondary effect in terms of the surface flow, reducing the distance for overland flow, and hence increasing the speed of the surface flow response.

Table 7.1 Average annual water balance for different stages of afforestation in the South Tyne (all values in mm)

Process	Baseline	Drained	Forested	50% Felled
Transpiration	126	143	166	74
Interception	159	159	440	287
Surface evaporation	144	112	81	148
Subsurface flow	374	951	669	557
Direct surface flow	725	145	148	429
Total runoff	1112	1105	823	997

Table 7.2 Flow statistics for different stages of afforestation in the South Tyne (all values in m^3/s)

Exceedance level	Baseline	Drained	Forested	50% Felled
0.1% exceedance	343	272	209	299
1% exceedance	94	88	61	79
10% exceedance	18	19	16	17
30% exceedance	8.2	9.3	6.9	7.5
50% exceedance	5.6	6.4	4.5	4.7
70% exceedance	4.2	3.3	2.5	3.3
90% exceedance	3.1	1.9	1.7	2.5
95% exceedance	2.7	1.7	1.5	2.3
99.9% exceedance	2.2	1.4	1.2	2.1

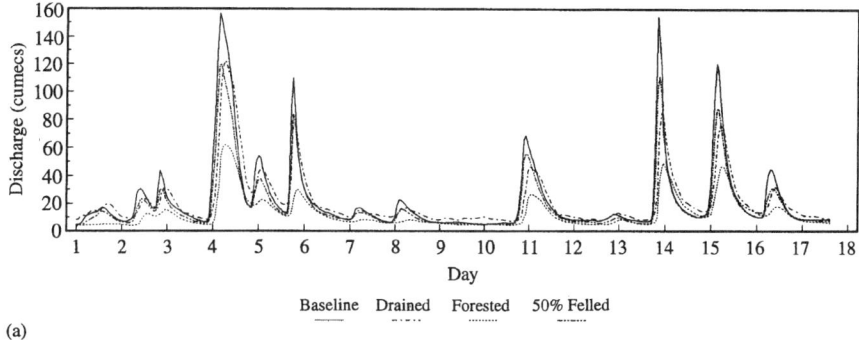

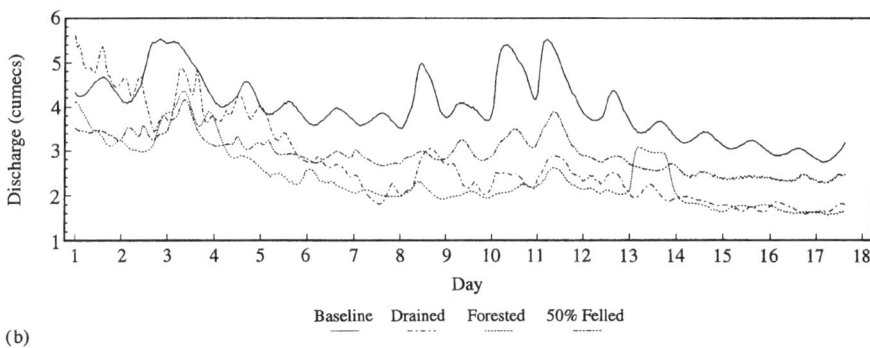

Figure 7.2 South Tyne: effect of afforestation on flows. (a) High flow period, August 1985; (b) low flow period, April 1985.

In the second simulation, the effect of the mature trees is to increase the amount of water evaporated from interception by the tree canopy, thus reducing the overall volume of runoff. The difference in interception loss is around 280 mm, which constitutes a decrease of 25% in the runoff. The effect of the ditches is lowered by their decrease in density and depth.

The third simulation shows that when felling of the forest begins there will be an increase in the total runoff volume. This occurs due to both the removal of the tree canopy, which decreases the interception loss, and the absence of any other vegetation cover, which removes the transpiration loss for the felled areas.

In summary, blanket afforestation of the South Tyne region could be expected to affect the hydrological regime of the area in a number of ways. Perhaps the most important of these would be the change in the

Table 7.3 Change in average probability of occurrence of breeding birds in the South Tyne under blanket afforestation

Species	Base line	Blanket afforestation
Coal tit	0.01	0.47
Crossbill	0.00	0.55
Golden plover	0.04	0.00
Lapwing	0.11	0.00
Red grouse	0.08	0.00

nature of the runoff regime from surface to subsurface, as a consequence of the intensive artificial drainage that would be necessary prior to afforestation. This is likely to affect both the levels of the peak flows and the baseflows in the streams. The reduction in runoff as a result of the increased interception loss from the forest canopy would be relatively insignificant, because of the high rainfall. Further increases in runoff might be expected at the end of the life cycle once felling of the trees begins.

(b) Ecology

The effect of blanket afforestation on ecology can be evaluated by comparing predictions of species distribution under the new land use, with the baseline conditions. In general, the variety of plant and invertebrate species is seen to decline throughout the area. The predicted distribution of birds changes dramatically with the disappearance of the populations of red grouse and golden plover. However, there would be a large increase in the distribution of the crossbill and other species associated with coniferous woodland such as the coal tit. Changes in the predicted bird population are summarized in Table 7.3.

(c) Economics

The economics of the afforestation proposal are dependent on the timescale of the required income from the land. In the short term, the forestry scheme would not be profitable in relation to other land uses. The land-owners would suffer a loss in income from sheep. The establishment of the forestry would therefore be strongly related to the level of grants available for planting and other incentives to encourage such forestry schemes.

(d) Summary of impacts

Blanket afforestation of the South Tyne region would cause significant changes to the species distribution of plants, invertebrates and birds. In particular, the predicted loss of the red grouse and the golden plover is a cause for concern. Changes to the volume of runoff from the region are relatively insignificant, but the predicted differences in baseflow might affect riverine ecology.

7.2.4 REFINEMENTS TO SOUTH TYNE FORESTRY PROPOSAL

In view of the undesirable impacts of the blanket coniferous afforestation, it is worth considering how certain refinements to the afforestation scenario might help to minimize these effects.

(a) Partial broadleaves

One possibility is to include some broadleaved species within the forest. Due to the nature of the climate, soils and topography the potential for growing broadleaved species is limited, but it should be possible to include species such as alder, birch and rowan to cover around 10% of the area.

The effects of this were assessed through additional model simulations. The results suggest that the modification would have only a marginal effect on the ecology. There would be some improvement in the status of the more generalist species, but the loss of the moorland species would still occur. In terms of the hydrology, the modification would have very little effect, although the winter loss of water from canopy interception would be reduced.

(b) Selective afforestation

A second possibility for minimizing the impact of the forestry would be to develop a more selective scheme, where only partial afforestation occurs and certain areas are left under their existing land use. The area to be left unaffected can be selected on the basis of environmental and economic criteria.

The location of the forestry is relatively insignificant in terms of the hydrology. The size of the impact will be approximately proportional to the afforested area. However, it is perhaps worth considering that the areas with the poorest quality soils will require the greatest land management, which in turn is likely to cause the greatest hydrological impact. In this case the poorest soils are the areas of Winter Hill peat on the hill tops.

From the ecological predictions, it can be seen that the most important change resulting from afforestation is the loss of the moorland species; in particular the golden plover and the cover types associated with the bogs and peatlands. The maps of predicted distribution for

Table 7.4 Effect of modified forest boundary on average probability of occurrence of breeding birds in the South Tyne

Species	Blanket afforestation	Modified forest
Coal tit	0.47	0.20
Crossbill	0.55	0.18
Golden plover	0.00	0.03
Lapwing	0.00	0.05
Red grouse	0.00	0.06

these species under the base-line conditions can be used to define the area that should be left unafforested on ecological grounds (Figure 7.1a–d).

The least economic areas for the forestry will be those regions where the climate and soils are poorest. Again this corresponds to the areas on the top of the hills, which are the most exposed, have the highest rainfall and have thick peat soils. The area can be identified from the area of Class 6.3 on the land capability map. This corresponds almost exactly to the area of the Winter Hill soil association. A second economic consideration is the profitability of areas which have a large population of red grouse. These too are found to occur on the high moorland.

From this discussion, it becomes apparent that the most appropriate areas to leave unafforested are the areas of moorland on the hill tops. This approach will satisfy each of the hydrological, ecological and economic criteria. A simple method of defining this area is to exclude all of the region above an elevation of around 450 m.

Applying the modified boundary of the afforestation to the models gives considerable improvement in the variety of species and plants and the effect on the predicted distribution of golden plover and red grouse. The changes to the predicted bird distribution are summarized in Table 7.4. The hydrological impact is also reduced as a result of the decrease in area that will undergo drainage and the decrease in canopy interception loss.

7.2.5 UPLAND FORESTRY CONCLUSIONS

The scenario of upland afforestation in the South Tyne suggests that a blanket afforestation scheme could have significant environmental impacts. The potential for different types of forestry in the area is limited by the physical conditions. However, a compromise solution of selective afforestation in the region should satisfy most requirements on hydrological, ecological and economic grounds.

7.3 Lowland forestry: the Cam Basin

The Cam Basin is a flat low-lying agricultural region in Eastern England. At present, apart from the city of Cambridge in the north of the catchment, the area is largely rural and the land is intensively managed for arable production. With the existence of large agricultural surpluses, a possible land use change scenario is that a community forest scheme might be set up in the region. The aim of this would be to improve the quality of the landscape, create new wildlife habitats and provide a recreational facility, as well as to provide an alternative source of income. The aim of this case study scenario is to identify an area, of around 10% of the catchment, that might be afforested, and to assess the environmental and economic consequences of this land-use change.

7.3.1 CAM BASE-LINE DATA

Precipitation in the Cam Basin is quite low, with an annual average of around 580 mm. The rainfall is distributed fairly uniformly throughout the year. The area is extremely flat in the northern half of the catchment, with the elevations ranging only from 10 to 50 m. There are slightly steeper slopes in the south where the land rises to 150 m.

The distribution of soils is quite complex. In general, the soils reflect the dominance of the underlying chalk geology, with a high percentage of the area having calcareous soils. Swaffham Prior is the main soil association found across the central belt of chalk, running from the SW to the NE. These are well-drained loamy brown calcareous earths. On the steeper ground of the chalk escarpment in the SW there is a block of Upton 1, which consists of shallow well-drained silty soils overlying the Middle Chalk. The clay loams of the Hanslope association are found on the gault clay in the NW and on the boulder clay in the SE. Soils along the river valleys are mainly loamy.

The influence of the chalk geology on the hydrology of the region is evident from the river network. There is no surface water system over the Middle Chalk, with springs emerging along the line of the base of the Middle Chalk outcrop. Much of the stream network in the NE of the catchment consists of large man-made drainage ditches.

The land in the Cam Basin is generally of very good agricultural quality, which explains the present dominance of arable activities, 86% of the area is of class 2 and a further 4% is of class 3.1 or higher.

In deciding the location for the new woodland and the type of forest appropriate for the Cam, a number of different options for forest mixes have been considered, corresponding to several of the options described by Dewar (1991). These are as follows:

- Spruce/Douglas fir plantations on better land in the lowlands;
- Pine in the lowlands;

- Seminatural broadleaves in the lowlands managed for non-market benefits;
- Non-native broadleaves on better land managed for timber production.

Given that the physical characteristics of the Cam Basin are quite extreme in terms of the dryness of the area and the chalk soils, it is important to consider the requirements of each of the forest types, before deciding the location of the forestry or its composition.

The first forest type consists of 35% sitka spruce, 35% Douglas fir, 10% larch, 10% broadleaves and 10% unstocked. Unfortunately, sitka spruce likes damp sites and grows best where there is more than 1000 mm annual rainfall. In addition, Douglas fir is unsuited to calcareous soils. Therefore, this mix of forest is considered inappropriate for the Cam Basin.

The second forest type of lowland pine is 80% Corsican pine, 10% broadleaves and 10% unstocked. Corsican pine likes light sandy soils and heavy clays in dry conditions. It is more successful than Scots pine on calcareous soils and as such it should be possible to cultivate in the Cam Basin.

The seminatural broadleaves would probably consist of a mix of species of oak, beech, ash and birch. All of these species like well-drained soil conditions, with particular preference for loams. Beech is the dominant species that is found on chalk soils in SE England, and ash also likes calcareous conditions. Thus the broadleaves would appear to be suited to the calcareous soils of the Cam Basin. The particular mix of broadleaves can be chosen to suit the site.

The forest type of non-native broadleaves managed for timber production is assumed to be 90% poplar and 10% unstocked. Poplar is demanding in terms of its location and requires loamy soils in sheltered locations. It grows well along stream banks. Certain locations within the Cam Basin should provide the appropriate conditions.

7.3.2 SELECTION OF REGION FOR CAM FORESTRY

All of the land in the Cam Basin is of reasonable quality for agricultural production. However, if an area of land is to be converted to forestry, then inevitably the poorest quality land will be chosen for this purpose. The quality of the land can be identified from the land capability map, from which it can be seen that approximately 10% of the area is of Class 3.2 or lower. Around 50% of this area forms a block of land on the chalk geology in the Royston area in the south west of the catchment. The remaining area lies mainly along the river valleys, with a scattering of areas in the east of the catchment. This seems to provide a reasonable distribution for the required level of afforestation, with a balance between small scattered areas of woodland and a few larger concentrated blocks.

The strip of land along the Rhee valley lies mainly on loamy soils such as Milton and Wantage 2. As such, this is an area which should be suitable for growing poplar. Most of the other areas lie on calcareous soils, which are best suited to seminatural broadleaves. However, it would be desirable if some areas of Corsican pine could also be included, as this would generate a suitable habitat for red squirrels. At present, there are no red squirrels in the Cambridgeshire area, with the nearest population in Thetford forest, but the squirrels could be reintroduced if the appropriate habitat was available. The proposal is therefore to divide up the assigned forestry area on calcareous soils to create 80% seminatural woodland and 20% Corsican pine woodland. To achieve this the existing distribution of deciduous and coniferous forest was used as a starting point. The surrounding areas were then buffered out, taking land from areas presently under arable cultivation, within the boundaries of the area defined by the poorest land capability. The resulting distribution of forest is illustrated in Figure 7.3(a).

7.3.3 CAM BASE-LINE MODELS

(a) Hydrology

The Cam Basin is predominantly a ground water system, with a large storage capacity for water within the chalk aquifer network. The response to rainfall events is generally slow and peak flows for the River Cam at Bottisham are around $25 m^3/s$. Since the annual rainfall is of the order of 580 mm and evapotranspiration rates average around 450 mm per year, the runoff ratio is extremely low at around 15%. The hydrology is also strongly influenced by management. Groundwater abstractions for public water supply amount to approximately 60 mm per year and around 35 mm per year is returned to the streams as sewage effluent. This accounts for a high percentage of the base flow in the streams during the summer. The net result is that there is little flexibility within the regional water balance for any increases in the amount of water use. The simulated water balance for the region as a whole is given in Table 7.5. With the knowledge that forestry frequently reduces the amount of available water, due to an increase in the loss from interception evaporation, it is important for this scenario to ensure that any changes to the regional water balance will be acceptable.

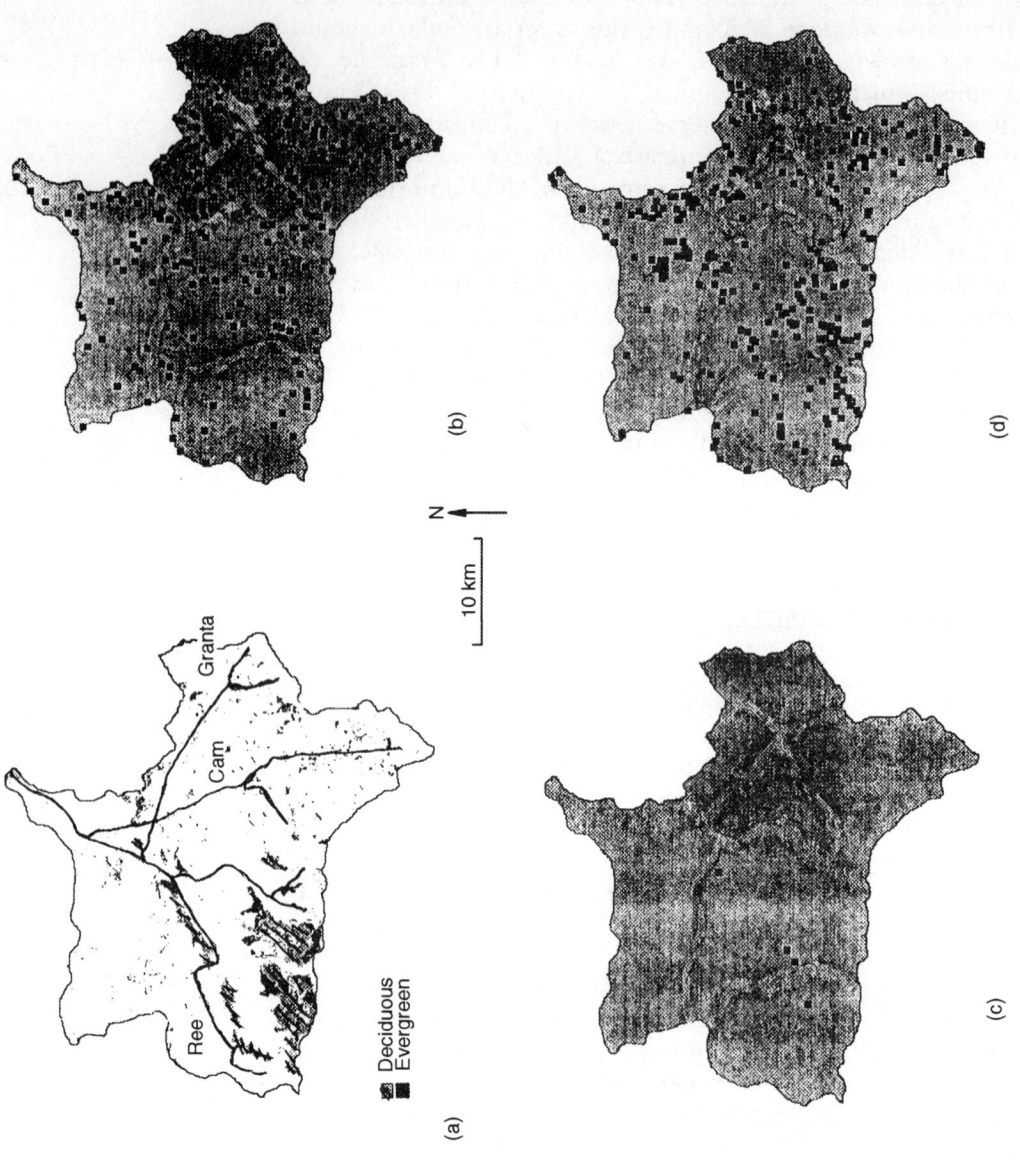

Figure 7.3 Cam: distribution of forestry and predicted distribution of squirrel populations. (a) Proposed distribution of Cam forestry; (b) grey squirrels before; (c) red squirrels after; (d) grey squirrels after.

Table 7.5 Simulated annual water balance for Cam catchment, 1989–93 (all values in mm)

Precipitation	560
Interception loss	69
Evapotranspiration	380
Groundwater recharge	96
Groundwater abstraction	58
Baseflow	43
Surface runoff	18
Surface water abstraction	1.7
Effluent return	38
Net channel flow	97

(b) Ecology

The ecological diversity of the Cam Basin with its existing land use is extremely poor, because of the intensive management for agricultural production. Greatest species abundance is found along the river corridors. Areas which are still traditionally managed provide habitats for some interesting inundation grasslands. Application of the spatial population dynamics model for squirrels, described in Chapter 4, indicates that there are no suitable habitats for the red squirrel. However, by contrast, the grey squirrel, which can subsist in suburban environments as well as deciduous woodland, is widely distributed. The predicted distribution of the grey squirrel is shown in Figure 7.3(b) for the existing land use. The intensively managed arable areas provide poor habitats for many species of breeding birds. Species characteristic of open farmland, such as the lapwing and skylark, are likely to occur in some of the grassland areas, but most woodland species are present in only very low numbers, or even absent, except in isolated copses (Table 7.6).

Table 7.6 Predicted probability of occurrence of birds in the mixed deciduous/coniferous forest of the Cam catchment

Species	*Base-line*	*Mixed forest*
Coal tit	0.00	0.40
Crossbill	0.00	0.12
Goldcrest	0.00	0.20
Lapwing	0.21	0.14
Long-eared owl	0.00	0.15
Siskin	0.00	0.16
Skylark	0.12	0.08

(c) Economics

The farm economics of the Cam Basin are clearly dominated by the income from arable activities. Any schemes aimed at encouraging diversification of land use will almost certainly be less profitable than arable activities, which means that if they are to be adopted in practice, adequate levels of subsidies or grants must be available to encourage uptake. There are a number of grants and schemes in existence to encourage forestry, but from the low take-up rate in good agricultural regions such as the Cam Basin, it would appear that these are insufficient to compensate for the loss of income. The main difficulty with forestry is that it is an investment, with substantial initial cost but no income return for as long as 50 years.

7.3.4 IMPACTS OF PROPOSED CAM AFFORESTATION

(a) Hydrology

The effects of the forestry scheme on the water balance were investigated at both the local and the catchment scales.

At the local scale, two SHETRAN plot models were used. The first plot model represents an area on the chalk soils around Royston, where a mix of Corsican pine and broadleaves is proposed. The second plot model represents an area of the proposed poplar plantation along the Rhee valley. Simulations were performed first with the present land use of arable cover and then converting this to both coniferous and deciduous tree cover.

The 'chalk' plot model represents an area where there are deep aquifers, to a depth of around 50 m, and the water table generally lies at a level of around 40 m below the ground surface. Because the water table is so deep, and the soils have a good infiltration capacity, the surface layers of the soil tend to become very dry. The ground slopes gently with a gradient of around 0.03. An area 500 m by 500 m is modelled using a grid resolution of 50 m. The boundary conditions for subsurface flow, in and out of the model, are estimated from the regional topography. The results for five-year simulations from 1989–1993 are summarized in Table 7.7. It can be seen that both deciduous and coniferous trees cause an increase in the interception loss. This leads to a reduction in the net rainfall, which in turn reduced the infiltration to the soil. However, the main effect of this is a reduction in the amount of water what is evaporated from the surface layers of the soil, which compensates to a large extent for the increase in interception loss. For arable land and deciduous trees the transpiration losses are very similar, but they are somewhat lower for coniferous

Table 7.7 Average annual water balance for Cam chalk plot model (all values in mm)

Process	Arable	Deciduous	Coniferous
Precipitation	575	575	575
Interception loss	75	209	296
Transpiration	170	182	148
Soil/surface water evaporation	174	92	82
Groundwater recharge	156	92	49

Table 7.8 Average annual water balance for Cam riverside plot model (all values in mm)

Process	Arable	Deciduous
Precipitation	575	575
Interception loss	73	202
Transpiration	294	295
Soil/surface water evaporation	182	91
Groundwater recharge	26	−13

trees. This is due to the decrease in soil moisture, which prevents the coniferous trees from transpiring at their maximum rate. In terms of the resulting recharge to the ground water, the deciduous trees cause a reduction of around 60 mm, whereas the coniferous trees cause a reduction of around 100 mm.

The riverside plot model used for the poplar plantation was identical to that used for the river corridor restoration study (Chapter 6). The water balance is summarized in Table 7.8 for arable and deciduous land covers. The interesting observation from these results is that the transpiration rates are considerably higher on this area than on the chalk, for both arable and deciduous covers. This suggests that each of the land uses suffers from some moisture stress in the chalk areas. Moisture stress is less of a problem in the riverside location as the level of the water table is much higher. However, as a result, the net recharge to the groundwater from both arable land and deciduous trees is negligible.

The conclusion from the small-scale plot models is that at the local scale, forestry on the area of chalk around Royston would cause a reduction in the groundwater recharge, but that forestry along the Rhee valley would have an insignificant impact.

The assessment of the hydrological impact was taken further by considering the overall effect on the catchment water balance, using

Table 7.9 Components of the average annual water balance at the subcatchment scale (all values in mm)

Simulation	Precip.	Intercep.	Evapo-Trans.	Base flow	Surface flow
Wimpole					
Baseline land use	572	68	401	49	26
After afforestation	572	99	377	46	23
Burnt Mills					
Baseline land use	559	67	373	32	8
After afforestation	559	90	356	30	7

the NUARNO model. The majority of the forestry is located in the subcatchments of the Rhee at Wimpole and the Rhee at Burnt Mills. Comparisons of the relevant components of the annual water balance for these subcatchments are given in Table 7.9. These figures show that the overall effect of the forestry is relatively small at the catchment scale. The majority of the increased interception loss takes place during the winter months and as such does not have an important effect on the critical low flows in the rivers during the summer.

(b) Ecology

The proposed expansion of forestry to include areas of coniferous woodland within a matrix of broadleaves creates the type of habitat with which the red squirrel is generally associated. Simulations with the population dynamics model suggest that there would be four areas where red squirrel populations could be maintained under the new land use (Figure 7.3c). Given the high degree of afforestation, this is a relatively low number of stable populations, which probably reflects the fact that the red squirrel requires relatively large areas of well-connected habitat to persist. Because the species occurs at relatively low densities, the fragmentation of coniferous habitats by deciduous woodland has been suggested as a factor threatening their persistence in landscapes (Gurnell, 1987). The population distribution of the grey squirrel would also increase as a result of the increase in forestry (Figure 7.3d). In particular an increase in distribution is predicted along the valley of the River Rhee.

The creation of a large area of mixed deciduous/coniferous forest in the south-west of the catchment will have a major impact on the breeding bird populations. Table 7.6 shows the predicted effects of such a land use change for seven species of lowland birds. Species characteristic of coniferous woodland, such as the coal tit, crossbill and goldcrest, are predicted to increase dramatically. Species found in

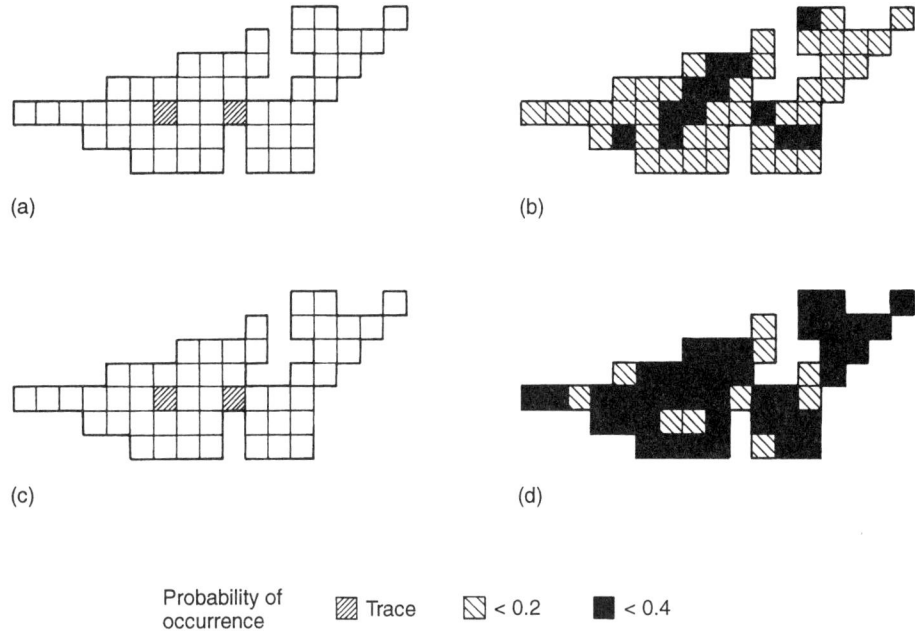

Figure 7.4 Predicted distribution of *Oxalis acetosella* (a, b) and *Primula vulgaris* (c, d) in the mixed deciduous/coniferous woodland. (a, c) Before forest; (b, d) after forest.

deciduous woodland, such as the long-eared owl, are also predicted to increase in number. The time taken for individual species to colonize the new habitats will vary, depending in particular on dispersal ability and breeding rates. The change in land use will lead to a decline in those species associated with open farmland, such as the lapwing and skylark. Large areas of the catchment, however, would be unaffected by the forestry schemes, and would therefore continue to provide suitable habitat for these species.

There is also the potential, with appropriate management, for a diverse flora to develop in the new lowland woodlands. Woodlands managed to create a mosaic of habitats, particularly those containing rides and glades, will contain more species than areas under blanket monoculture afforestation. Such woods will also, of course, provide more habitats for invertebrates, in addition to plants, birds and mammals. Figure 7.4 shows the predicted increase in the distribution of two species of woodland plants, *Oxalis acetosella* and *Primula vulgaris*. Such species will probably slowly disperse into the new deciduous woodland from the small patches of extant deciduous woodland around which the new woodland is planted.

Table 7.10 Marginal benefits and grants for different forest types

Forest type	Marginal benefit (£/ha)	Deciduous (£/ha)
Broadleaved	−271	54
Conifer	−242	13
Poplar	338	0

(c) Economics

Costs and returns for each of the forest types described by Dewar (1991) were included in the catchment economics LP. The impact of the forestry could then be investigated by changing the land use from arable to each of the forestry types over the appropriate areas of each land capability class. Model runs were compared for the Cam Basin with the existing arable land use and with the proposed forested area. These showed that the loss of income resulting from the forestry scheme would amount to £427 000 per annum. This is equivalent to £56/ha.

Under the Farm Woodland Scheme (ATP-Landbase, 1994), the grant that would be available for planting this type of forest is estimated at £39/ha, when annualized at a rate of 6%, over a 50-year period. Although not quite sufficient to compensate for the loss of income of £56/ha, this does not appear too unreasonable. However, when the results from the LP are analysed in more detail, looking at the different forest types individually, it is apparent that it is not this simple. Table 7.10 gives the marginal benefits and grants that are available for the three forest types. This shows that the grants available for planting both broadleaves and conifers are not nearly sufficient to compensate for the costs involved in this change of land use. The reason that the overall figures appear to be reasonable is due to the profitability of poplar plantations.

7.3.5 LOWLAND FORESTRY CONCLUSIONS

The proposed afforestation scheme would provide several, new, varied and attractive landscapes, which would provide a recreational facility for the population of the Cambridgeshire area. The analyses with the NELUP models indicate that the forestry should enhance the variety of wildlife in the region and should not have a serious hydrological impact. However, the economics model suggests that the present grant schemes are inadequate to compensate for loss of income for planting either broadleaves or conifers. A new initiative would be necessary to persuade landowners to implement the proposed change in land use.

References

ATP-Landbase (1994) *Reaping the Benefits*. ATP-Landbase, Penrith.

Countryside Commission National Forestry Development Team (1993) The national forest strategy: draft for consultation, Cheltenham, 88pp.

Dewar, J. (1991) Forestry expansion – a study of technical, economic and ecological factors. New planting methods, costs and returns. Forestry Commission Occasional Paper no. 46.

Dunn, S.M. and Mackay, R. (1996) Modelling the impacts of open ditch drainage on hydrology. *Journal of Hydrology*, (in press).

Gurnell, J. (1987) *The Natural History of Squirrels*, Helm, London, 201pp.

8 The marginal value and hydrological impact of agricultural irrigation: a representative farm-level case study for the Cam river catchment

8.1 Introduction

The Catchment Area of the River Cam is defined by the National Rivers Authority (NRA) as an area of approximately 110000 ha, mainly in Cambridgeshire, but also incorporating parts of Bedfordshire, Essex, Hertfordshire and Suffolk. Within the Cam Catchment, NRA data suggest that only 1% of annual ground water supplies are shown to be used for agricultural purposes. Taken as an annual figure, this seems low and suggests that water authorities should be little concerned about the withdrawals made for agriculture. However, given the dominance of arable farming within the region and the relative unimportance of livestock demands on water supplies, it is likely that the agricultural demands on water resources will only be significant within the dry periods of the year. Given this seasonal nature of agricultural demands for water, they could account for a much larger proportion of water withdrawals during certain months and hence be a more important target for water conservation policy.

Within the region covered by the Farm Business Survey (FBS) for the Eastern Counties of England (Murphy, 1994), an area wholly encompassing the catchment as defined by the NRA, 28% of farms use some level of irrigation as part of their cropping system. The total area of land actually irrigated within the FBS region occupies approximately 7% of total agricultural land (Murphy, 1993). Of this irrigated land, although the data provided by the FBS do not tell us for which particular crops the irrigation is used, it can be deduced that approximately 90% of the land irrigated is *associated* with farms which produce potatoes, sugar beet or both.

More specifically, during the harvest year 1991/2, 59 FBS farms were found to operate an irrigation system *and* carry out production of potatoes, sugar beet or a combination of the two. Paring the area of land down to these crops and the area irrigated for each of the 59 farms allowed a *t*-test* of the 59 paired observations to be carried out. The results of this test showed no significant difference (actual *t*-statistic = −1.033, critical *t*-statistic = ±2.00) between the area down to these crops and the area irrigated, across the whole sample. Using this evidence, it can be deduced that the principal crops which utilize irrigation systems within the Cam catchment are potatoes and sugar beet. Consequently, this study focuses on farms producing only these two crops.

8.2 Modelling approach

Across the whole FBS sample (including farms with and without irrigation systems) the average area sown to potatoes in 1991/2 was 30 ha per farm, whereas the average area sown to sugar beet was 41.3 ha (Murphy, 1993). Using these areas as representative of a potato and sugar beet farm within the Cam river catchment, and assuming no other cropping activities, the first part of this study uses a farm-level Linear Programming (LP) model to investigate the proportionate increase in yields under irrigation required to make an irrigation system economically viable at current depreciation and running costs (AGRO, 1992) and water abstraction charges (Nix, 1992). This is done by comparison of the marginal value product (shadow price) of irrigated (MVP_i) and non-irrigated (MVP_n) land.

These marginal values tell us the additional income that is attainable through the employment of one extra hectare of potato and sugar beet production, and are given in the LP output as the dual activity under each land constraint. Each marginal value is calculated on the basis that labour and machinery use, land rentals and overheads would increase with the expansion of land use, but assume that fixed machinery stocks on the farm would not have to be expanded to incorporate the additional production; it is assumed that existing machinery is not working at full capacity. The point at which it becomes more economically viable to irrigate the potato and sugar beet enterprise is where irrigated yields are sufficiently higher than non-irrigated yields to make MVP_i greater than MVP_n. As the results show, the proportionate

---

* Hypothesis: H0: Area (Potatoes/Sugar Beet/Both) for farm *i* = Area Irrigated for farm *i*
  H1: Area (Potatoes/Sugar Beet/Both) for farm *i* ≠ Area Irrigated for farm *i*

increase in irrigated yields actually required is estimated at 23% over non-irrigated yields.

The values used as the base yield for each crop (41.2 tonnes potatoes/ha, 49.9 tonnes sugar beet/ha) were generated using the crop growth model, EPIC (Jones et al., 1991), under a non-irrigated system for MLURI land class 2 (Bibby et al., 1991). The rationale behind focusing on this grade of land is simply because it is the dominant land classification found within the Cam catchment, representing approximately 78% of total agricultural land. In addition, nitrogen fertilizer application rates were fixed at 200 kg N/ha for both crops and were assumed to remain constant under all situations.

The second part of this study assumes that the yield levels attainable under irrigation are fixed. Data supplied by an independent field trial, involving 23 farms in East Anglia (Hinton and Varvarigos, 1990), revealed the proportionate increases in yields that are attainable under irrigation systems for various crops. Specifically, the report outlined an average yield response to irrigation (in a dry year) of approximately 2.0 tonnes/ha/25 mm of water applied for potatoes, and 3.3 tonnes/ha/25 mm of water applied for sugar beet. Using suggested application rates (144 mm/year for potatoes and 120 mm/year for sugar beet) this meant that irrigated potato and sugar beet yields could be fixed at approximately 12 and 16 tonnes/ha, respectively, above base levels.

Applying these absolute yield increases to the base yields estimated by EPIC (used in the first part of the study) potato yields are found to be 29% higher under an irrigated system, whereas sugar beet yields are found to be 33% higher. When applying the results of part one, this suggests that MVP_i for the whole farm plan will indeed be greater than MVP_n, and hence an irrigated system under these base conditions would be more profitable than a non-irrigated one. What this second section does is to analyse by how much current irrigation running costs and water abstraction charges would have to be increased for MVP_i to fall below MVP_n, assuming these observed irrigated yield levels can be attained. In this way, the study estimates a possible 'irrigation abatement tax' that could be levied on producers in order to make irrigation economically non-viable and hence, assuming profit maximization objectives, reduce the use of water supplies for agricultural purposes.

The third part of the study demonstrates a link that can be made between economic and hydrological modelling systems. Once the farm-level economic model has been run to suggest the possible magnitude of an irrigation abatement tax, the NELUP hydrological modelling system can be used to assess the impact that this reduction in irrigation would have on net runoff.

Although the licensed irrigation abstraction levels indicate that irrigation is an insignificant component of the water balance when averaged over the entire catchment and the full year, it is likely that there is some local effect on the water balance in those areas where irrigation occurs. Thus, a decrease in irrigation could be valuable if it helps to increase low flows which have been found to reach critical levels in small streams during the summer. This issue is tackled by a hydrological analysis of the effects of changes in irrigation, using the water application rates from the farm-level economic model, for the same representative farm in the Cam catchment.

The farm-level economic model used was representative of an 'average' potato and sugar beet farm in the Cam catchment. This involved taking average FBS recorded requirements for labour, machinery and capital; the average rentals charged and overheads involved in such a farm situation; and the average prices received for the produce (Murphy, 1994). The objective function of the model was to maximize farm management and investment income – net farm income minus an imputed charge for the manual labour provided by the farmer and spouse. Separate activities were specified within the model regarding irrigated and non-irrigated potato and sugar beet production. This was done in order to specify the differences in labour and machinery requirements under irrigated and non-irrigated production (AGRO, 1992) and to allow the yields attained under an irrigated system to vary for both crops.

8.3 The marginal value product of irrigated land across increasing yields

Once established, the farm-level economic model was run to represent the average base situation of a farm producing 30 ha of potatoes (limited by area quota) and 41.3 ha of sugar beet on non-irrigated land. Under this scenario, the model produced a farm management and investment income of £22 509, generating a value for MVP_n of £165/ha. Explicitly, this implies that an additional hectare available for potato or sugar beet production, on non-irrigated land, would increase farm management and investment income by £165, assuming land and property charges were increased proportionately. The corresponding value for the same farm plan using FBS reported net margins per hectare at 1991/2 prices (Murphy, 1993) is £358, but is taken before land and property charges. Adding land and property charges back into the model solution, MVP_n is equal to £382/ha, which suggests the model to be fairly accurate.

In order to determine MVP_i for potatoes and sugar beet at various yields, the yields attained on irrigated land were allowed to vary in 5% increments from the yields attained under non-irrigation, as estimated by EPIC, up to 30% above those yields. Clearly, because irrigation involves positive additional costs (AGRO, 1992), MVP_i will lie below

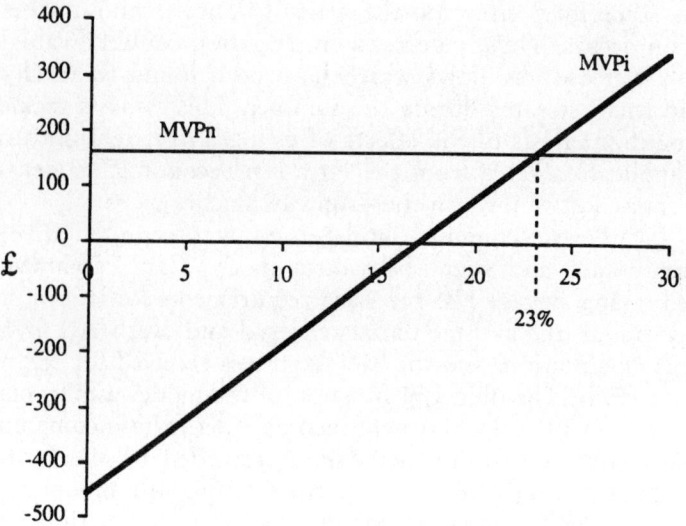

Figure 8.1 Increasing irrigated yields above non-irrigated yields.

MVP_n when yields are identical. The object is to find the percentage increase in combined potato and sugar beet yields at which MVP_i becomes equal to MVP_n.

Figure 8.1 shows how MVP_i for the farm plan increases as yields under irrigation increase, and illustrates the yield level above which using an irrigated system becomes more profitable. As the graph shows, the irrigated system becomes more economically viable when combined potato and sugar beet yields rise 23% above the yields attained under non-irrigation.

The yield levels observed by Hinton and Varvarigos (1990) suggest that potato yields will actually be 29% higher under an irrigated system, whereas the comparable figure for sugar beet is 33% higher. By applying these increased rates to the farm plan used in this study, the area-weighted average increase in yields between the two crops under irrigation is found to be 30.7%. Using the values generated for Figure 8.1, projection of this average increased yield level against the MVP lines tells us that at this point MVP_i lies 120% above MVP_n, at £363.

These results suggest that irrigation is actually more profitable under these base conditions on the representative farm. Presuming that farmers operate with profit maximizing objectives, irrigation will be widely used among potato and sugar beet producers.

If water authorities *were* interested in limiting the use of water for irrigation purposes, one possible means of achieving that objective would be to increase the cost to the farmer of running an irrigation system (possibly through restricting the rate at which water can be abstracted) or by directly increasing water abstraction charges. Simple application of the above results might suggest that this tax level would simply need to be equal to the difference between the two MVPs at the observed irrigated yield levels. This value is fairly high, at approximately £198/ha, and represents an increase of 57% on current running and abstraction costs.

However, rather than considering a pooled increase in yield as above, we need to consider separate increases in yields for each of the two crops, specifically, an increase of 29% for potatoes and 33% for sugar beet. Also, because additional irrigation running costs would also imply additional labour, machinery and capital charges, knock-on effects occur within the farm system making the calculation of such a tax more convoluted. Furthermore, with higher yield levels under irrigation, relative harvesting times are likely to increase and must be incorporated into the analysis. These complexities necessitate the use of the farm LP model and suggest that an increase in costs lower than 57% should be expected.

8.4 Increasing irrigation costs to non-viable levels

Using the same representative farm LP model, this section illustrates by how much the total irrigation charges would have to increase in order for MVP_i to again equate with MVP_n, given that the potato and sugar beet yields observed by Hinton and Varvarigos (1990) under irrigation are attained. Figure 8.2 represents the model output when combined irrigation running costs and abstraction charges are increased in 10% increments.

The outcome of the analysis suggests that in order for the MVP of irrigated land to fall below that of non-irrigated land, then total irrigation costs per hectare would have to increase by 26%. Based on current irrigation costs (AGRO, 1992; Nix, 1992) in order to achieve this outcome, and force irrigation out of a profitable farm plan, the actual cost per hectare would have to rise by £90. This figure is lower than initially anticipated in Section 8.3 (£198), but still represents a sizeable tariff to have to impose on farmers. Alternatively, this figure could be viewed as the average or representative amount of compensation the 'representative' potato and sugar beet farmer would require per hectare in order to be persuaded to forego irrigation.

This analysis, however, was carried out assuming that prices received for irrigated and non-irrigated produce would be the same. In practice, the price that crops such as potato and sugar beet can command depends critically on product quality. In simple terms, the quality of

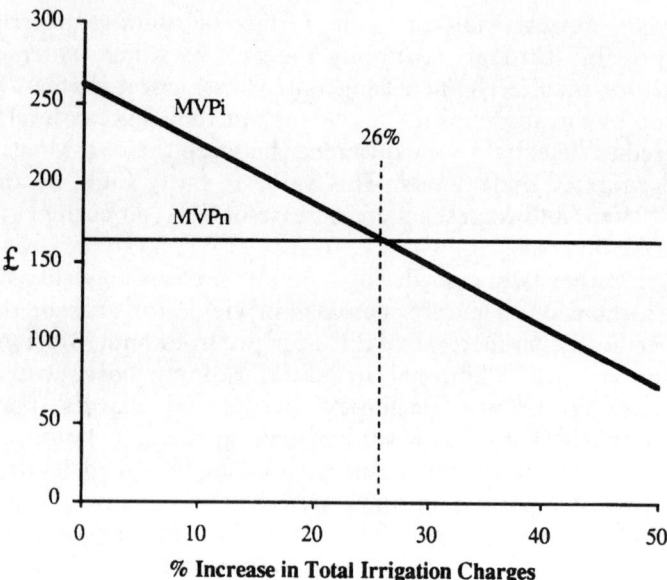

Figure 8.2 Increasing irrigation charges to non-viable levels.

potatoes and sugar beet is positively related to the water content of the product which, in the case of sugar beet, has further consequences for sugar content. Clearly, if price advantage was available for irrigated produce, then the estimated increase in costs required to force irrigation out of an economically viable system would be higher. With this in mind, a sensitivity analysis of increasing price commanded by irrigated produce is relevant.

Because these crops reach high yields per hectare, small increases in price per tonne will lead to large absolute increases in gross crop revenue. Therefore, the increase in irrigation costs required for MVP_i to equal MVP_n can be expected to be particularly sensitive to any change in price. Figure 8.3 demonstrates a model parameterization of irrigated potato and sugar beet price, from base levels (where irrigated price equals non-irrigated price) to irrigated product prices lying 10% above non-irrigated prices. These price levels are mapped against the corresponding increase in irrigation costs that would be required to force irrigation out of an economically viable system.

Figure 8.3 shows, for example, that if irrigated product prices commanded a 5% advantage, the corresponding increase in irrigation costs required to make irrigation non-viable (i.e. at $MVP_i = MVP_n$) would be 71%, or an increase in costs of £246/ha. Clearly, if this was the case, then the imposition of an irrigation abatement tax could become more difficult and expensive to police.

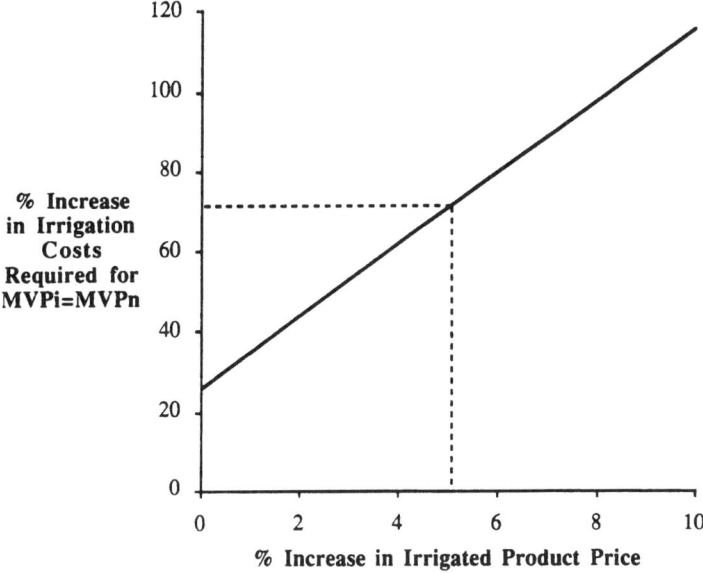

Figure 8.3 The effects of an irrigated product price advantage.

8.5 The hydrological impact of changes in irrigation practice

The hydrological modelling in this section was performed using the plot model described in Chapter 3, representative of an area on the chalk region of the Cam catchment. The analysis performed in Chapter 3 indicated that this was a region where irrigation is likely to be used, because most crops would suffer moisture stress without it. For this analysis, four management scenarios were considered; irrigated potatoes, non-irrigated potatoes, irrigated sugar beet and non-irrigated sugar beet. Following the management assumed within the farm-level economic model, the irrigation of potatoes consisted of an annual application of 144 mm/ha, whereas the irrigation of sugar beet consisted of an annual application of 120 mm.

The hydrological issue of interest relates to the effect of irrigation on the water balance. Table 8.1 compares the average annual water balance for each simulation. These results show that of the 144 mm of irrigation water applied to potatoes, some 82 mm is used by the crop to increase transpiration and a further 9 mm is an additional interception loss. Only 53 mm of the irrigation water is recycled back into the groundwater to return eventually to the stream. Thus, there is a net loss from the stream of 91 mm. Similarly, for sugar beet there is a net loss of

Table 8.1 Average annual water balance for irrigated and non-irrigated potatoes and sugar beet (all values in mm)

Process	Irrigated potatoes	irrigated sugar beet	Non-irrigated potatoes and sugar beet
Rainfall	575	575	575
Irrigation	144	120	0
Interception loss	84	83	75
Transpiration	251	232	169
Soil evaporation	174	174	174
Groundwater recharge	210	206	157

71 mm from the stream. The abstractions from the stream take place during the summer when flows are at their lowest.

Predictions from the NUARNO model show that the summer base flow in the Cam at Bottisham is typically around 1–1.5 m^3/s and, of this, around two thirds is generated by effluent return. This means that the equivalent depth of runoff, contributing to the catchment base flow during the summer, is of the order of only 1.5 mm per month. In other words, the net water use for irrigation in June and July is around 20 times the average amount of water that is locally supplied to the stream. If a number of farms in a particular area were all using irrigation on their fields, this could have a significant local impact on the level of water in the streams during the summer months. This aspect is disguised by the totals for the catchment, because only a small proportion of all farms grow crops requiring irrigation.

8.6 Conclusions

This study demonstrates a method by which a farm-level LP model can be used at a representative scale to assess the likely costs of a certain hypothesized policy instrument. If water authorities were concerned that agricultural irrigation was making excessive demands on water supplies, then this method could be used to estimate the likely costs of restricting water supplies to the average irrigated farm. The study also shows how the model can be adjusted for different estimates of price advantage from irrigated systems so that the user can assess the costs likely to arise across different farm situations where product quality is variable.

Further, the link between the economic and the hydrological modelling systems shows that if the levels of irrigation were to be reduced by the imposition of an irrigation tax, the reduction in water use could lead to a significant improvement in the levels of flow in small streams during the summer, in those areas where irrigation is used.

References

AGRO Business Consultants Ltd (1992) *Farm Machinery Costs*, 3rd edn, AGRO Business Consultants Ltd, Melton Mowbray.

Bibby, J.S., Douglas, H.A., Thomasson, A.J. and Robertson, J.S. (1991) *Land Capability Classification for Agriculture*. Macaulay Land Use Research Institute, Aberdeen.

Hinton, L. and Varvarigos, P. (1990) *Observations on the Economic Performance of Irrigation – A Study of 23 Farms in East Anglia*. Agricultural Economics Unit, Department of Land Economy, University of Cambridge, Cambridge.

Jones, C.A., Dyke, P.T., Williams, J.R. *et al.* (1991) EPIC: an operational model for evaluation of agricultural sustainability. *Agricultural Systems*, 37, 341–50.

Murphy, M.C. (1993) *Report on Farming in the Eastern Counties of England, 1991/92*. Agricultural Economics Unit, Department of Land Economy, University of Cambridge, Cambridge.

Murphy, M.C. (1994) *Report on Farming in the Eastern Counties of England, 1992/93*. Agricultural Economics Unit, Department of Land Economy, University of Cambridge, Cambridge.

Nix, J. (1992) *Farm Management Pocketbook*, 22nd edn. Wye College, University of London.

9 Discussion

The overall objectives of the NERC/ESRC Land Use Programme were clearly defined from the outset, but it is only with hindsight that it is possible to appreciate fully how much needs to be invested in a multidisciplinary team. While it is exciting to hear a problem discussed from the angle of a discipline other than one's own, there are barriers to understanding, in the form of concepts, jargon, assumptions and working methods. It took almost a year of discussing land use in weekly seminars with each group seeing it from the perspective of its own discipline, before each group had the confidence to criticize the proposals put forward by the others. Multidisciplinary approaches, especially quantitative ones, are dogged by questions of complexity, scale and data. Confidence within NELUP grew as every member of the team began to use the same datasets, especially the remote sensed land cover data. The literature of land-use studies, much of which is descriptive, was not as useful as might have been expected. Prior to 1980 welfare economics exerted a strong influence on how planning should be approached, while ecology and hydrology contributed to the research from within their independent disciplinary interests. Post 1980, the dominance of the 'market' became more overt, moderated, to a growing extent, by concern for the environment.

Six key issues related to land use planning and management have been identified during the research programme.

1. **Digital technology.** New technology for acquiring, managing and analysing information offers possibilities for tackling the complexities of land use in a realistic way. As the research progressed, it became obvious that information about spatial problems, and the analyses of spatial data, can often be communicated more appropriately through a computer screen than through the printed word. Without geographic information systems, it is extremely difficult to construct and manage an adequate database for spatial problems. It is now clear why much of the older literature on land-use made a distinction between 'place' and 'space'. It was either forced to present information in detail about small areas, which emphasized the variations between sites, or to describe in a broad way what happened in 'representative areas' where variation was lost in the averaging. Digital technology can handle spatial

problems in a spatial way, and display variation without losing the continuities that are found in the landscape.

2. **Systems modelling**. Land use is a subject which embraces a wide range of functions and interests. As a resource for the production of food and fibre and water supplies, aspects of the subject have been studied in detail over many years. Much of the process understanding has been the fruits of reductionist science conducted in such different disciplines as physics, chemistry and biology. Attainment of a successful crop production cycle lies in the combination of groups of processes into a coherent whole. When processes are combined, there are inevitably interactions between them. Systems modelling is complementary to process modelling because it is only at the systems level that the objectives of complete production cycles can be defined, that the process studies can be integrated, and the interactions between different processes properly appreciated.

Systems modelling, especially in its quantitative form, can bring together parallel consideration of the processes and their interactions. It is, for example, a good way of aggregating pollution loads from different parts of a system and of estimating the total impact they make on the common resources in the public domain: air quality, water resources, biodiversity and fragile ecosystems. At present the amount of knowledge available on processes is usually far greater than on the systems in which they are active. However, by using well-researched interpretations of the component processes as building-blocks for a systems model, the generalist planner/manager can explore the interactions, especially when there is a Decision Support System, which allows hypothetical experiments to be conducted on the system. Where there are potential conflicts of interest about land use between the holders of the property-rights and other stakeholders with an interest in the countryside, systems modelling can bring transparency to the debate and help in classifying the trade-offs that may be possible.

3. **Scale**. There is no single spatial or temporal scale appropriate to land use studies. At one end of the spectrum are the evolutionary processes, which are measured in geological scales of space and time, and at the other extreme are local perturbations. The majority of the available data are the fruits of research that favours controlled experiments, which tend to be constrained by cost, with the result that the experiments are conducted at relatively minute scales. The scales that are relevant to systems modelling can be discussed against the background of the problems to be examined.

In the real world, the majority of land use questions fall in the broad area of policy:

- Much of management is devoted to implementing policies in the most effective way in order to meet their objectives, which have been agreed long ago;
- Another set of questions is posed by policy failures, which call for interventions that are needed to correct unwelcome side-effects of policies which have been implemented earlier;
- The third and most interesting set of questions can be categorized as planning problems, where the likely outcomes of policies need to be appraised ahead of the decision on implementation. Proper analysis at this stage should help to define the management tasks, estimate the likely impacts of the policy on those directly affected by it as well as on the wider environment, and reduce the likelihood of policy failures.

Clearly, there is a strong political dimension to land-use studies, which in turn is a clue to the appropriate scales. In terms of time, concern is likely to be focused on the very short-term in correcting policies, which have been implemented but are perceived to be politically unacceptable. The long-term is probably 10–15 years, which is devoted to the broadly specified strategic problems which opinion-formers have signalled to be important. Sustainability, climate change, environmental impacts, biodiversity are examples of such problems. Past-experience in dealing with land-use change shows that change is likely to be gradual, to be achieved by modifying existing policies, and to have a high social and political content.

The areal scales also reflect the boundaries of political influence. In land use the supernational policies of EU exert a major influence. The whole country is the domain of national policies, which are handed down, occasionally with regional modification, through the countries and administrative districts.

Scale raises several issues to be addressed by the land-use modeller. Models for strategic and operational purposes have different specifications whereas the data supply places limitations on the information that is available for model building. The NELUP response is that the methodology of systems analysis in land use is robust and takes cognisance of scale. It can be applied within different boundaries as required. Stretching of a model by aggregation from one scale to a much larger one is not recommended, particularly as models built at different scales are likely to be required for different purposes. Data should be matched to models whose complexity is tailored to a particular set of requirements, an approach that is demonstrated by treating farms as nested within the catchment to which they belong.

4. **Biophysical/economic sciences.** When land is used for growing a crop, two distinct sets of processes occur, one of which falls within biophysics and the other within economics.

The crop production process is composed entirely of biological/physical/chemical processes, which explain the mechanisms of crop development from germination to harvest, the demands for water and nutrients, the pathology of pests and disease, and the energy transformations. Crop-growth is an input–output transformation process of long duration, driven by solar energy, where the attainment of a successful product, in terms of both quantity and quality, depends on the delivery of the correct mix of inputs at the right time. Scientists usually interpret the biophysical model of production in quantitative terms of yield per unit area, without paying too much attention to the costs of the inputs necessary to produce that yield.

The farmer converts the biotransformation process of crop production into a business by assigning prices both to the product which is marketed, and to the resources used to produce it. In effect, the biophysical production process becomes a value-added exercise in which prices are fixed either by the market or by government. Success in business depends on maximization of the value-added exercise, which in the simplest case is the difference between the price of the output and the costs of the inputs used to attain that output.

Economics complements biophysics in the analysis of land use problems. However, full complementation will only be achieved when economics can include properly the costs of the externalities of a production process in the value-added equation.

5. **Regulation.** Where the market is perceived to be incapable of setting prices, which would lead to an acceptable equilibrium between supply and demand or of dealing with the externalities of the production process in the form of pollution, other instruments are used in an attempt to regulate the market and deliver what society needs. There is a long tradition of intervention by Governments in the markets for agricultural produce with the objectives of boosting farm incomes, setting quality standards or encouraging selected forms of production. In the EU, we have price supports, quotas and set-aside, all designed with the purpose of regulating production.

The long-term impact of regulation on producers is debatable. In agriculture, regulation insulates producers from the markets and reduces their sensitivity to the changing needs of consumers. EU policies favour a small number of commodities to the point where they lead to monocultures and crowd-out almost every other crop production possibility that does not carry a subsidy. Regulations are blunt instruments for allocating land between different uses, because they fail to reflect, and in some cases even try to compensate for, the most basic constraint on the biophysical production process, land capability. NELUP results emphasize the range of land capabilities that need to be recognized, sometimes even in relatively small areas.

As a way of fostering more interesting, mosaic-type landscapes, spatial diversity of land capability points towards diversity of land uses.

'Command and control' strategies for the regulation of pollution have been shown to be economically inefficient. More responsive approaches, based on markets, need reliable information flows of the kind found in a systems network and lead to better integration of the management of land, air and water.

6. Integration of modelling, data collection and monitoring. NELUP has been an exercise in applied research, which has demonstrated the feasibility of a systems approach to the study and management of land and water. To a large extent, it has been a retrospective exercise in that the modelling framework was developed on the basis of information that was already available in the public domain. Only a very small amount of survey work was carried out in the Tyne catchment in order to check the accuracy of the remote sensed land cover map. Many assumptions were made about how 'reality' could be simplified, while still retaining an adequate understanding of the principal processes which occur in land use. Two major simplifications were: treating the models as deterministic, and ignoring the transient conditions in moving from one steady state to another. Confidence in the system is based on the close agreement between the estimates from the models and the historical records of land uses, river discharges and ecological surveys.

Although NELUP is no more than a first stage in demonstrating how decision-making in the complex allocation of spatial resources may be improved through using the digital technology, it indicates the way in which data-collection, modelling and decision-support tools can be integrated. The proper starting-point, for any organization which wishes to follow this path, is agreement about the level of modelling that is needed in order to represent their objectives in managing resources. Only when the data requirements of the models have been defined is it possible to make an informed decision about data collection, especially on what are the absolute minimum requirements and how often do they need to be updated. Whenever possible, data should be collected in digital form and handled automatically. Too often, programmes start as an ambitious data-collection exercise, and later wonder what to do with the information. Data collection is usually very expensive.

The great advantage of organizing decision-making around a model is that it provides a logical framework within which to formulate the problem, it helps to control the acquisition of data and it provides a method for the transparent analysis and comparison of alternatives. When a decision has been made to implement a policy, monitoring of its objectives should be linked to the model. In this way, learning from the system becomes continuing and the results of the monitoring exercise become a time-series of performance which can be used to update and refine the model.

Appendix: The catchments of the Rivers Tyne and Cam

River Tyne

The catchment of the River Tyne in north-east England (Figure A.1) is 2935 km^2 in area and contains a population of over 810 000. The major population centres are at Newcastle and Gateshead in the east of the catchment, with smaller towns such as Hexham and Haltwhistle further upstream to the west. The catchment ranges in altitude from sea level to 893 m; the two main tributaries of the River Tyne, the River North Tyne and River South Tyne, rise in the hills of the Cheviots and North Pennines, respectively, and have their confluence about 5 km west of Hexham. The geology is varied, with Carboniferous limestone, Millstone Grit and sandstone in the North Pennines and Cheviots, and coal measures underlying in the Newcastle area. Annual rainfall varies from less than 700 mm near Newcastle, to over 1800 mm at Kielder Forest.

There is a wide variety of land use within the catchment; sheep rearing is dominant on the rough grazing and heather moorland in the upper reaches of the catchment, with dairy farming being practised at medium to lower altitudes. Arable farming is confined to the catchment downstream of the North Tyne/South Tyne confluence, eventually reaching the urban/industrial conurbations of Newcastle and Gateshead. Most of Kielder Forest (60 000 ha) falls within the catchment, the coniferous forest forming a dominant feature of the landscape around the River North Tyne. Heather moorland is most abundant in the south of the catchment, with large areas used for grouse shooting.

In terms of nature conservation, the catchment contains 77 Sites of Special Scientific Interest (SSSI), including three National Nature Reserves (NNR). Most of the northern part of the catchment is within the Northumberland National Park, whereas the North Pennines Area of Outstanding Natural Beauty contains much of the River South Tyne and its tributaries. In addition, part of the Pennine Dales Environmentally Sensitive Area includes the Rivers South Tyne, West Allen and East Allen.

194   Appendix: catchments of the Rivers Tyne and Cam

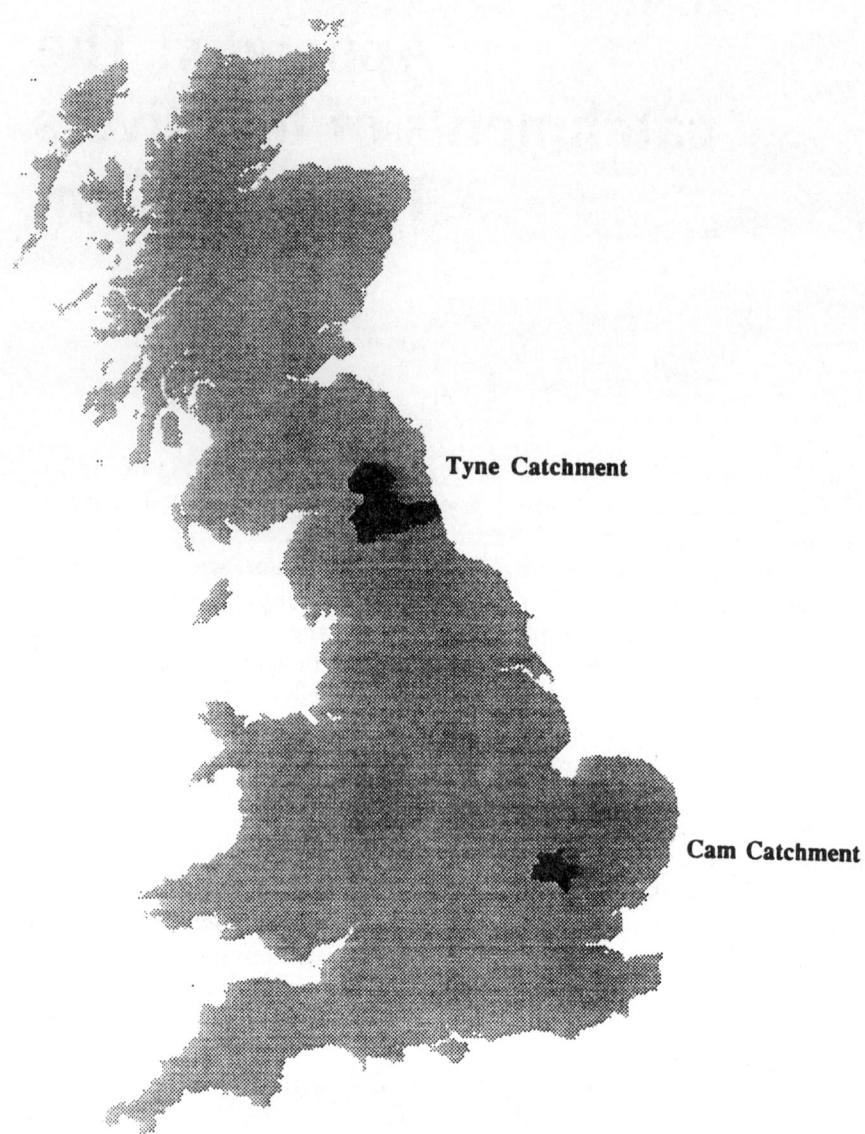

Figure A.1 Catchments of the Rivers Tyne and Cam.

River Cam

The catchment of the River Cam in south-east England (Figure A.1) is approximately 1100 km^2 in area, and contains a population of 250 000. The major population centres are Cambridge, Royston and Saffron Walden. The three main tributaries, the River Granta, River Rhee and Bourn Brook, flow into the River Cam about 5 km upstream of Cambridge. Geologically the catchment splits into three around Cambridge: to the West, clay overlies Lower Greensand; to the South and

East, chalk is overlain by boulder clay; and to the North peat fenland is dominant. The altitude ranges from sea level in the fens to 122 m in the chalk uplands. The rainfall is also low, varying from 500 mm over parts of the fens, to 650 mm in the chalk 'uplands'.

The dominant land use in the catchment is arable agriculture, with cereals (especially winter wheat), oilseeds and rootcrops dominant. Grassland is restricted to only around 30 000 ha, mainly permanent pasture with some rotational leys.

The catchment contains 40 SSSIs, 23 of which depend on the continuation of a suitable water regime. The largest of these, Wicken Fen, is managed by the National Trust, but the Fen is now more than 2 m above the surrounding land due to shrinkage, and water levels are maintained by pumping. Cambridgeshire County Council has designated four 'Areas of Special Importance for Nature Conservation' within the catchment: the South Cambridgeshire Chalk, the Cam Washlands, the Granta Valley, and the area of clay in the south west of Cambridgeshire. There are no National Parks or ESAs within the catchment.

Index

Page references in **bold** refer to figures; those in *italics* refer to tables

AGENDA 21 4
Aggregate-level LP model 32, 33–4, 35, 57
Agricultural irrigation, effect on Cam river catchment 178–86
　hydrological impact 185–6
　increasing costs to non-viable levels 183–4
　marginal value product across increasing yields 181–3
　modelling approach 179–81
Agricultural land capability classification 30, 31, 89, 122
Agriculture Act (1986) 71
Analogue model 112
Arable Area Payments 133, 134
ARCINFO 121
Areas of Outstanding Natural Beauty 3
ARNO model 46, **47**
　data sets *61*
Assemblages 91

Bayesian theory 19, 31, 77
Behavioural models 97
Biodiversity 13
Biological function values 6
Biophysical processes 7–14, 190–1
Boussinesq equation for two-dimensional groundwater flow 52
Buffer zones 137, 145–6

Calluna vulgaris 93–5, 97, 100, 160
　distribution **95**
　incidence *94*
Cam, River, catchment 194–5, **194**
　geology **60**
　monthly runoff and flow variance **59**
　see also Deintensification of agriculture; Forestry
Carbon–oxygen cycle 6, 9–10
Catchment-level land-use modelling 32–4, *35*
Classification algorithms 91–2
Common Agricultural Policy 132
Communities 91
Community Forests 156
Compaction techniques 117
Computer systems 15–16
Countryside Commission 33, 132, 156
Countryside Survey (1990) 17, 88

Data structures 114
Database management systems 114–15
Decision making, different approaches 110–12, **111**
Decision support systems (DSS) 16, 114–15
　see also NERC/ESRC Land Use Programme
Deintensification of agriculture 132–53
　land use policy and 152–3
　in Pennine Dales ESA 146–52
　　birds **150–1**, *151*
　　ecological effects of ESA prescriptions 149–51, *149*
　　economic effects of ESA prescriptions 148
　　grasslands management 147–8
　　implications of ESA policy 152
　　plants 150–1
　in River Cam catchment 134–46
　　amenity usage 135–6
　　defining the river corridors 137–9
　　ecological, economic and hydrological management and predictions 139–46
　　ecology of the river corridors 139–42
　　economic effects of river corridor habitat management 143
　　effects on river water quality 143–6, **144**
　　hydrological protection 136–7
　　nature conservation 136
　　riparian birds 141
　　rotational set-aside, environmental effects 134
　　targeted, semipermanent

set-aside in lowland river corridors 135–7
terrestrial vegetation 139–41
Denitrification 137, 145–6
Deterministic modelling approaches 83–6
Detrended Correspondence Analysis 80
Development Plan System 3
Diffuse-source pollution 137
Digital technology 188
Direct use values 6
Discriminant analysis 77–8
District-Wide Local Plan 3
DYNUT 49

Ecological function values 6
Ecological model
  forestry 86–90
  of land use 19–20, 122–3
  landscape 159–60, 169–71
Ecology 13–14
Economic model of land use 18–19, 24–39, 122, 128-9
  agricultural land capability classification 30
  analytical economic framework 24–6
  choice of empirical modelling technique in NELUP 26–9
  components of models 29
  data collation 30–2
  financial land-use measures 37
  forestry 160, 172
  model construction 32–4
  model usage 36–7
  physical land-use measures 37
  profit-maximizing behaviour 37–8
    alternatives to 37–9
  validation 34–6
English Nature 3
Environmentally Sensitive Areas 71, 132, 133–4, 153
Erosion Productivity Impact Calculator (EPIC) 31, 32, 35, 53
ESRC 86
European Union (EU) 4, 73, 132
  Drinking Water Directive 137
  Fifth Action Programme 4
Existence value 7

Farm Business Survey (FBS) 19, 30, 31, 32, 35, 121, 178
Farm Woodland Scheme 176
Farm-level LP modelling 32, 179–86
Fertilizers 1, 74–5
Forestry 156–76
  Cam Basin
    base-line data 167–8
    base-line models 169–72
      ecology 169–71
      economics 172
      hydrology 169
    impacts of proposed afforestation 172–6
      ecology 174–5
      economics 176
      hydrology 172–4
    selection of region 168–9
  South Tyne 158–76
    base-line data 158–9
    base-line models 159–60
      ecology 159–60
      economics 160
      hydrology 159
    impacts of afforestation 160–5
      ecology 164
      economics 164
      hydrology 160–4
      summary of 164–5
    refinements to proposal 165–6
      partial broadleaves 165
      selective afforestation 165–6
  use of DSS in case studies 157–8
Forestry Commission 156

Generalized Additive Models (GAMs) 81
Generalized linear-modelling (GLIM) 79, 81
Geographic information systems (GIS) 15, 20, 30, 31, 76, 97, 115–18
Goodness of fit 113
Graphical user interface (GUI) 16, 128
GRASS 121, 139
Grasshopper warbler 141, **142**

Habitat distribution 20
Heather, see *Calluna vulgaris*
Hierarchical structures 115
Hierarchy of values 5
Hydrological cycle 6, 7–9, 43–5, **44**
Hydrological modelling 19, 42–69, 123
  applied to Cam river basin 58, **59**
  database 60–2
  detailed analysis 64–6
  forestry 159, 169
  hydrological processes 43–5
  modelling process 57–60
  modelling systems 45–57
  parameter uncertainty 68
  scale 67–8, 189-90
  scenario analysis 62–4
  see also NUARNO; SHETRAN

Iconic model 112
Institute of Hydrology (IH) 60
Institute of Terrestrial Ecology (ITE) 17, 19, 89, 93, 96
Intrinsic values 6

Lagged explanatory models 82
Lagged response models 82
Land classification 19, 89–90, 96
  see also Agricultural land capability classification
Land Cover Map of Great Britain 17, 18, 19

Land use policies, trends in 2–7
Landsat satellite imagery 17, 57, 89, 92
Landscape complementation 96
Landscape ecology 71–105
  application of associative models 90–101
    species utilizing more than one habitat 96–101, **98**
    species utilizing one or more habitat states exclusively 91–6
  application of deterministic models 103, **104**
  associative models
    aspatial, for predicting species distribution 76–81
    spatial 81–3
  developing integrated modelling system to evaluate ecological changes 86–90
  ecological processes 72–3
  geophysical and geochemical processes 72–3
  human–economic processes 72–3
  investigating 75
  processes, scale and 72–5
Landscape Response Units 73
Lapwing 97, 151
Lin's formula 49
Linear programming (LP) model 19, 27–9
Logistic regression 79

Macaulay Land Use Research Institute (MLURI) 30, 157
  agricultural land capability classification 30, 31, 89, 121
MacSharry reforms 132
Mahalanobis distance 78
Markov modelling 84
Mathematical modelling 112–13, **113**
Matrix modelling 19–20, 96

Metapopulation dynamics theory 83, 84
Ministry of Agriculture, Fisheries and Food (MAFF) 4, 61, 132
  compensation payments 133
  Farm Business Survey 19
  Farm Census (June) 18, 122
  Habitats Scheme 133
Multicriteria decision making (MCDM) 38
Multicriteria programming (MCP) 38–9

National Parks 3
National Vegetation Classification 20, 91, 93, 98, 100, 140
Nature Conservancy Council 88, 91, 93
  Phase I information 92
  Phase I methodology 88–9, 92
  Phase 1 Vegetation Survey 18
Neighbourhood models 85
NERC 86
NERC/ESRC Land Use Programme (NELUP) 14–18, 86
  data flows **21**
  database 20–1
  decision support system 16, 22–3, **23**, 118–20
    assessing usability of 128–9
    building 127–8
    database 119–22, *120*
    interface design 123–5
    models 18–20, 121–3, 127–8
      associative 122–3
      ecological 19–20, 122
      economics 18–19, **29**, 121–2, 128
      farm-level 122
      hydrological 19, 123
      process oriented 122–3
      regional 122
    using 126–7

  land cover triangle 16–18, **17**
    agricultural census data 18
    field surveys of habitats 18
    remote sensed land cover map 17–19
  methodology 16
  screen display 21–2
Network structures 115
Nitrate concentrations 45, 52, 57, *145*
Nitrate Sensitive Areas 132, 133, 137
Nitrogen 12
Nitrogen cycle 10–11, **11**
Nitrogen modelling system (NMS) 53, **56**
NUARNO model 19, 42–3, 46–52, 58, 123, 128
  flow 46–9
  flow equations used in *48*
  predicted flows for Cam catchment **63**
  scale and 67–8
  scenario analysis 62–4
  transport 49–52
  transport equations *50–1*
Nutrient cycles 11–13

Option values 6
Oracle 121
Ordinary least squares (OLS) 31
Ordination techniques 80
Ordnance Survey Digital Terrain Model 60

Penman–Monteith interception model 47
Pennine Dales Environmentally Sensitive Area 71
Percentage absolute deviation (PAD) 36
Pesticides 74–5
Photosynthesis 11
Planning and Compensation Act (1991) 3
Point-source pollution 137
Pollution 1, 6, 137
Population dynamics GIS

modelling, combined approach 102, **103**
Population dynamics models 83–6
Principal Components Analysis (PCA) 80
Principle of parsimony 87
Profit maximization 37–8, 39

QUICKSORT 52

Raster data structures 117–18
Real-world systems 112–14
Reciprocal Averaging (RA) ordination 80
Red grouse 97, **99**
Regional Planning Guidance 3
Regression modelling techniques 78, 81
Regulation 191–2
Relational data structures 115
Relational database management system (RDBMS) 121
Remote sensing 15
Remote-Sensed Classification of Land Cover 23
Richard's equation 52
Rutter interception model 47

SAS 29

Satellite imagery 89, 96
  Landsat 17, 57, 89, 92
Scale, significance of 190
Set-aside 82, 133, 134, 135–7
SHE-SHELL 62
SHETRAN 42–3, 46, 52–7, 58, 123, 128, 143
  contaminant transport 56–7
  data sets 61
  flow 52–3, **53**
  flow equations 54–5
  local model **65**
  scale and 67–8
  transport 53–7
Soil Survey of England and Wales 60
Solar energy 7
Spaghetti data structure 116
Spatial analysis 117
Spatial decision support systems (SDSS) 115
Spatial organization 14
Spatially lagged models 81, 82
Structure Plans 3
Structured query language (SQL) 120
Système Hydrologique Européen (SHE) hydrological project 19, 21, 46, 123
Systems modelling 189

Tessarae 73
Tessellation data structures 117
Tessellations 85
This Common Inheritance 1
Topological data structures 116
Total economic value 5–7
Tyne catchment 193–4, **194**
  see also Forestry

UNIX 29
User values 6, 7

Vector data structures 116–17

Water systems, importance of 4–5
Watershed storage capacity curve (WSCC) 47
Weighting matrix 82
Within-cover change model for plant communities 20

X11 protocol 128
Xview library 128

Yellow wagtail 141, **142**
Yields 1, 121

DATE DUE

JUN 27 1997	

UPI 261-2505 G   PRINTED IN U.S.A.